Real Analysis:
A Historical Approach

Real Analysis: A Historical Approach

Saul Stahl

A Wiley-Interscience Publication
JOHN WILEY & SONS, INC.
New York / Chichester / Weinheim / Brisbane / Singapore / Toronto

Copyright © 1999 by John Wiley & Sons, Inc. All rights reserved.

Published simultaneously in Canada.

For ordering and customer service, call 1-800-CALL-WILEY.

Library of Congress Cataloging in Publication Data:

Stahl, Saul.
    Real analysis : a historical approach / Saul Stahl.
       p. cm.
    " A Wiley-Interscience publication."
    Includes bibliographical references and index.
    ISBN 0-471-31852-3 (cloth : alk. paper)
    1. Mathematical analysis.   2. Functions of real variables.
 I. Title.
QA300.S882   1999
515—dc21
                                       99-21917
                                         CIP

Printed in the United States of America

10 9 8 7 6 5 4 3 2 1

Preface

A Focused and Historical Approach

The need for rigor in analysis is often presented as an end in itself. Historically, however, the excitement and impatience that characterized the first century and a half of calculus and its applications induced mathematicians to place this issue on the proverbial back burner while they explored new territories. Ironically, it was developments in physics, especially the study of sound and heat, that brought the pathological behavior of trigonometric series into the foreground and forced the mathematical world to pay closer attention to foundational issues.

It is the purpose of this text to provide a picture of analysis that reflects this evolution. Rigor is therefore introduced as an explanation of the convergence of series in general and of the puzzling behavior of trigonometric series in particular.

The first third of this book describes the utility of infinite, power, and trigonometric series in both pure and applied mathematics through several snapshots from the works of Archimedes, Fermat, Newton, and Euler offering glimpses of the Greeks' method of exhaustion, preNewtonian calculus, Newton's concerns, and Euler's miraculously effective, though often logically unsound, mathematical wizardry. The infinite geometric progression is the scarlet thread that unifies Chapters 1 to 5 wherein the nondifferentiability of Euler's Trigonometric Series provides the crucial counterexample that clarifies the need for this careful examination of the foundations of calculus.

Chapters 6 to 10 consist of a fairly conventional discussion of various aspects of the completeness of the real number system. These culminate in Cauchy's criterion for the convergence of infinite series which is in turn applied to both power and trigonometric series. Sequential continuity and differentiability are discussed in Chapters 11 and 12 as is the maximum principle for continuous functions and the mean value theorem for differentiable ones. Chapter 13 covers a discussion of uniform convergence proving the basic theorems on the continuity, integrability, and differentiability of uniformly convergent series and applying them to both power and trigonometric series. While the exposition here does not follow the historical method, a considerable amount of discussion and quoted material is included to shed some light on the concerns of the mathematicians who developed the key concepts and on the difficulties they faced.

Chapter 14, The Vindication, uses the tools developed in Chapters 6 to 13 to prove the validity of most of the methods and results of Newton and Euler that were described in the motivational chapters.

Intended Audience for this Book

This text and its approach have been used in a Junior/Senior-level college introductory analysis course with a class size between 20 and 50 students. Roughly 40% of these students were mathematics majors, of whom about one in four planned to go on to graduate work in mathematics; another 40% were prospective high school teachers. The balance of the students came from a variety of disciplines, including business, economics, and biology.

Pedagogy

Experience indicates that about three quarters of the material in this text can be covered by the above described audience in a one-semester college-level course. In the context of this fairly standard time constraint, a trade-off between the historical background material and the more theoretical last two chapters needs to be made. One strategy favors covering uniform convergence (through Cauchy's theorem about the continuity of uniformly convergent series of continuous functions (13.2.1–3)) instead of the more traditional Maximum Principle and Mean Value Theorem.

Each section is followed by its own set of exercises that vary from the routine to the challenging. Hints or solutions are given for most of the odd-numbered exercises. Each chapter concludes with a summary. The excerpts in the appendices can be used as independent reading or as an in-class line-by-line reading with commentary by the professor.

This book contains sections which can be considered optional material for a college course; these sections are indicated by an asterisk. The least optional of the optional sections is 4.2 which describes the rudiments of Newton's Polygon Method. While the details look daunting, the material can be covered in two 50-minute long lectures. The purpose of Section 6.4 is to provide an elementary and self-contained proof of the existence of irrational numbers. Section 9.4 is included as an explanation of some concerns of higher mathematics. All of these optional sections reinforce the importance of infinite series and infinite processes in mathematics.

The Riemann integral proved to be a thorny problem. Within the limitations of one semester it is impossible to do justice to historical motivation, the traditional contents of the course, uniform convergence, and the Riemann integral. Therefore I have decided to state and make use of the integrability of continuous functions without proof. This is applied for the first time in the proof of Theorem 13.2.4, a place to which a typical class is unlikely to get in one semester.

SAUL STAHL

Lawrence, Kansas

Acknowledgments

Fred van Vleck proofread portions of the manuscript. I am indebted to him and to my colleagues Bill Paschke and Pawel Szeptycki for their patience when they were subjected to my ramblings and half baked ideas. My thanks are also due to Jessica Downey, Andrew Prince, and Sharon Liu of John Wiley & Sons, Inc. and to Larisa Martin and Sandra Reed for their help in converting my notes into a book.

Your comments are welcome at Stahl@math.ukans.edu

S. S.

Contents

*This book is dedicated to
the memory of my parents and my brother,
Finkla, Moses, and Dan Stahl*

1

Archimedes and the Parabola

How did Archimedes evaluate the area of the parabolic segment almost 2,000 years before the birth of calculus?

1.1 THE AREA OF THE PARABOLIC SEGMENT

One of the most important issues considered in the context of elementary geometry is that of the area of polygonal regions. The only curvilinear figure commonly studied at this elementary level is the circle, and once again its area is a major concern, as are its circumference and the construction of tangents. It is only natural to extend these notions to other curvilinear figures. Analytic geometry and calculus were developed, at least in part, in order to facilitate investigations of this advanced level.

The ancient Greeks, in whose culture modern science and mathematics are rooted, developed their own version of calculus in order to compute areas and volumes. This Greek calculus, which they named *the method of exhaustion,* was invented by Eudoxus (408?–355 B.C.) and perfected by Archimedes (287–212 B.C.). The latter actually had several different techniques, not all of which bear a resemblance to the methods used in today's calculus texts. Only one of Archimedes's accomplishments is discussed here, and that in summary form—the determination of the area of the region bounded by a parabolic arc and its chord. Such details as Archimedes supplied appear in Appendix A and are given some justification in Section 1.2. The present section is limited to a discussion of those aspects of Archimedes's arguments that are germane to the subsequent development of calculus.

The parabola is one of several curves, collectively known as the *Conic Sections*, that were first studied in depth by Menaechmus (circa 350 B.C.) of the Platonic school. The book *Conic Sections* written by Archimedes's contemporary Apollonius of Perga (circa 262–circa 190 B.C.) contained over 400 propositions about these curves.

The *parabola* was originally defined as the section of a right circular cone made by a plane that is parallel to one of the cone's generating lines (Fig. 1.1).

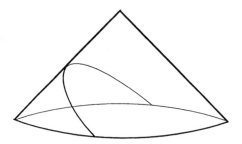

Fig. 1.1   The parabola as a conic section

Figure 1.2 illustrates definitions needed to describe Archimedes's argument. Given any two points Q and Q' on a parabola, the portion of the parabola

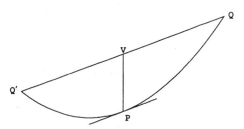

Fig. 1.2   A typical parabolic segment

bounded by them is an *arc*, and the straight line segment QQ' is its *chord*. The region bounded by an arc of the parabola and its chord is called a *segment* (of the parabola) and the chord is said to be the *base* of this segment. Given any segment of a parabola, that point P on the arc at which the tangent is parallel to the base is called the *vertex* of the parabola (see Exercise 1.2.3).

The crux of Archimedes's argument is the following geometrical proposition (see Fig. 1.3). Here and elsewhere we adopt Euclid's convention that the equality (or inequality) of geometrical figures refers to their areas only. Thus, the triangle's median divides it into two equal (though in general not congruent) portions.

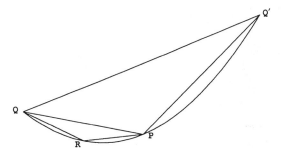

Fig. 1.3 Triangles parabolic segments

Proposition 1.1.1 *Let QQ' be the base of a parabolic segment with vertex P. If R is the vertex of the parabolic segment with base PQ, then*

$$\Delta PQQ' = 8\Delta PRQ.$$

Archimedes's proof of this fact appears both in the next section and in Appendix A. Here we are concerned only with how this proposition can be applied to the evaluation of the area of the parabolic segment. Archimedes's idea was to use this proposition to obtain successive approximations of the area of the parabolic segment by polygons that consist of conglomerations of inscribed triangles. For the parabolic arc QQ' with vertex P of Figure 1.4, the first approximating polygon $\mathbb{P}_1$ is $\Delta PQQ'$. If R and R' are the vertices

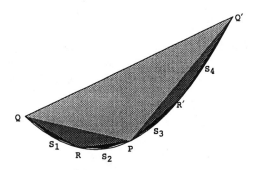

Fig. 1.4 Approximating a parabolic arc

of the parabolic arcs with bases PQ and PQ' respectively, then the second approximating polygon $\mathbb{P}_2$ is the pentagon $QRPR'Q'$ obtained by augmenting $\mathbb{P}_1$ with ΔPQR and $\Delta PQ'R'$. Since each of these augmenting triangles has area one eighth that of $\Delta PQQ'$, it follows that

$$\mathbb{P}_2 = \mathbb{P}_1 + \frac{1}{4}\mathbb{P}_1.$$

If S_1, S_2, S_3, S_4 denote the vertices of the segments with bases QR, RP, PR', $R'Q'$ respectively; then the third approximating polygon $\mathbb{P}_3$ is obtained by augmenting $\mathbb{P}_2$ with ΔQS_1R, ΔRS_2P, $\Delta PS_3R'$, $\Delta R'S_4Q'$, each of which has area

$$\frac{1}{8}\Delta PQR = \frac{1}{8}\Delta PQ'R' = \frac{1}{8}\cdot\frac{1}{8}\Delta PQQ'.$$

Thus,

$$\mathbb{P}_3 = \mathbb{P}_2 + 4\cdot\frac{1}{64}\mathbb{P}_1 = \mathbb{P}_1 + \frac{1}{4}\mathbb{P}_1 + \frac{1}{16}\mathbb{P}_1.$$

In general, each iteration of this process adjoins to each triangle Δ of the previous iteration two new triangles whose total area equals one fourth that of Δ. Hence the total area of the chain of triangles adjoined in one iteration is also one fourth the total area of the triangles adjoined in the previous iteration. Consequently,

$$\mathbb{P}_{k+1} = \mathbb{P}_1 + \frac{1}{4}\mathbb{P}_1 + \cdots + \frac{1}{4^{k-1}}\mathbb{P}_1 + \frac{1}{4^k}\mathbb{P}_1 = \left(1 + \frac{1}{4} + \cdots + \frac{1}{4^k}\right)\mathbb{P}_1.$$

Archimedes next argued that the parabolic segment is indeed the limit of the polygon $\mathbb{P}_k$. This he justified by pointing out that in Fig 1.5, where P is

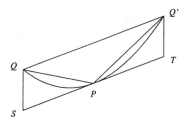

Fig. 1.5 A triangle inside a parabolic segment

the vertex of the parabolic arc QQ', and $\mathbb{PS}$ is an abbreviation for *parabolic segment* PQQ',

$$\mathbb{PS} < \text{parallelogram } QSTQ' = 2\Delta PQQ'$$

$$\therefore \quad \frac{1}{2}\mathbb{PS} < \Delta PQQ'$$

$$\therefore \quad \mathbb{PS} - \Delta PQQ' < \frac{1}{2}\mathbb{PS}.$$

Thus, each iteration of the approximating process reduces the uncovered portion of the parabolic segment by more than half. Formally,

$$0 < \mathbb{PS} - \mathbb{P}_{k+1} < \frac{1}{2}(\mathbb{PS} - \mathbb{P}_k).$$

Is this sufficient reason to guarantee that the difference between the parabolic segment and the approximating polygons vanishes in the limit? Archimedes showed great perspicuity by recognizing that there was a deep principle involved in this question which merited explicit formulation as a new axiom which now bears his name. This axiom is mentioned again in Chapter 6 where it is replaced by the Completeness Axiom. Archimedes accepted this argument as sufficient grounds to conclude that indeed

$$\text{parabolic segment } PQQ' = \text{limiting position of } \mathbb{P}_k.$$

Finally, Archimedes computed the limiting value of the area of the approximating polygon $\mathbb{P}_k$. This computation made use of an analog of the well known formula for the sum of a finite geometric progression. As this formula will be used repeatedly in the sequel, it is stated here in two equivalent forms and given the status of a proposition. The straightforward proofs are relegated to Exercise 10.

Proposition 1.1.2 *If n is a non negative integer, then*

1. $\quad b^n - a^n = (b - a)(b^{n-1} + ab^{n-2} + a^2 b^{n-3} + \cdots + a^{n-2}b + a^{n-1})$

2. $\quad 1 + a + a^2 + \cdots + a^{n-1} = \dfrac{1 - a^n}{1 - a} \qquad \text{if } a \neq 1. \hfill (1)$

Archimedes employed this proposition together with a rather awkward procedure involving a proof by contradiction, to demonstrate a special case of the *infinite geometric progression*'s formula

$$1 + a + a^2 + a^3 + \cdots = \frac{1}{1 - a} \qquad \text{if } -1 < a < 1. \tag{2}$$

Equation (2) marks the first appearance of the infinite geometric progression in this text. We shall return to it repeatedly in the sequel and it will be fully justified in Proposition 9.1.1. From basic calculus, it is known that Equation (2) follows from Equation (1) because for $-1 < a < 1$ the limiting value of a^n in Equation (1) is 0. Accordingly,

$$1 + \frac{1}{4} + \frac{1}{4^2} + \frac{1}{4^3} + \cdots = \frac{1}{1 - \dfrac{1}{4}} = \frac{4}{3}$$

and so

$$\mathbb{PS} = \lim_{k \to \infty} \mathbb{P}_k = \lim_{k \to \infty} \left(1 + \frac{1}{4} + \cdots + \frac{1}{4^k}\right) \mathbb{P}_1$$

$$= \left(1 + \frac{1}{4} + \cdots + \frac{1}{4^k} + \cdots\right) \Delta PQQ' = \frac{4}{3} \Delta PQQ'.$$

Archimedes's main conclusion is now stated explicitly.

Theorem 1.1.3 *Let P be the vertex of the parabolic segment with base QQ'. Then*

$$\text{area of segment } PQQ' = \frac{4}{3} \Delta PQQ'.$$

Exercises 1.1

1. Show that

a) $1 + \dfrac{1}{4} + \dfrac{1}{16} + \dfrac{1}{64} + \cdots = \dfrac{4}{3}$

b) $1 - \dfrac{1}{4} + \dfrac{1}{16} - \dfrac{1}{64} + \cdots = \dfrac{4}{5}$

c) $1 + \dfrac{3}{5} + \dfrac{9}{25} + \dfrac{27}{125} + \cdots = \dfrac{5}{2}$

d) $3 - \dfrac{9}{4} + \dfrac{27}{16} - \dfrac{81}{64} + \cdots = \dfrac{12}{7}$

e) $1 + \dfrac{1}{4} + \dfrac{1}{16} + \cdots + \dfrac{1}{4^8} = \dfrac{87381}{65536} = 1.333328\ldots$

f) $1 - \dfrac{1}{4} + \dfrac{1}{16} - \cdots + \dfrac{1}{4^8} = \dfrac{52429}{65536} = .800003\ldots$

g) $1 + \dfrac{3}{5} + \dfrac{9}{25} + \cdots + \left(\dfrac{3}{5}\right)^8 = \dfrac{966721}{390625} = 2.474805\ldots$

2. Let $\angle ABC$ have measure α and let the points $A_1, A_2, A_3, \ldots$ be defined as follows:

$$A_1 = A,$$
A_2 is the foot of the perpendicular from A_1 to BC,
A_3 is the foot of the perpendicular from A_2 to AB,
A_4 is the foot of the perpendicular from A_3 to BC,
A_5 is the foot of the perpendicular from A_4 to AB,

$\ldots$

Show that

a) $A_1 A_2 + A_2 A_3 + A_3 A_4 + \cdots = \dfrac{A_1 A_2}{1 - \cos \alpha}$

b) $A_1 B + A_3 B + A_5 B + \cdots = A_1 B \csc^2 \alpha.$

3. The tortoise challenges the swift footed Achilles to a race. "However", says the tortoise, "since you are such a good runner, I demand a 1000 ft lead." Achilles accepts, whereupon the tortoise says to him: "You might as well concede the race now, for I will prove to you that you can never catch up with me. For the sake of the argument let us say that you are ten times faster than I. By the time you have reached my starting place, I will be 100 ft ahead of you. By the time you have reached that spot, I will be 10 ft ahead of you. By the time you reach that spot, I will be 1 ft ahead. By the time

you have covered that foot, I will still be one tenth of a foot ahead of you. Thus, you see, I will always be ever so slightly ahead of you. So why don't spare yourself the effort and just pay me off right now." Explain the fallacy in the tortoise's reasoning. Assuming that Achilles runs at the rate of 30 ft/sec, and granting the tortoise his stipulated ratio of 10 to 1, where and when will Achilles overtake the tortoise?

4. The middle third of the unit interval $[0, 1]$ is removed. Next, the middle thirds of each of the remaining two intervals are removed. At the third stage the middle third of each of the remaining four intervals is removed and so on. Compute the sum of the lengths of all the extracted intervals.

5. The rightmost half of the unit interval $[0, 1]$ is removed. Next, the rightmost third of the remaining interval is removed. At the third stage the rightmost quarter of the remaining interval is removed and so on. Compute the sum of the lengths of all the extracted intervals.

6. The rightmost third of the unit interval $[0, 1]$ is removed. Next, the rightmost third of the remaining interval is removed. At the third stage the rightmost third of the remaining interval is removed and so on. Compute the sum of the lengths of all the extracted intervals.

7. Let n be any integer ≥ 2. The rightmost n-th part of the unit interval $[0, 1]$ is removed. Next, the rightmost n-th part of the remaining interval is removed. At the third stage the rightmost n-th part of the remaining interval is removed and so on. Compute the sum of the lengths of all the extracted intervals.

8. The decimal number $.a_1 \ldots a_s a_{s+1} \ldots a_t a_{s+1} \ldots a_t a_{s+1} \ldots$ is denoted by $.a_1 \ldots a_s \overline{a_{s+1} \ldots a_t}$. Express the following decimal numbers as the ratio of two integers.

a) $.\overline{5}$ b) $.0\overline{5}$ c) $.12\overline{7}$

d) $.\overline{27}$ e) $.35\overline{27}$ f) $.\overline{731}$

g) $.000\overline{731}$ h) $.\overline{1234}$ i) $.123\overline{1234}$

9. Prove that every decimal number of the form $N.a_1 \ldots a_s \overline{a_{s+1} \ldots a_t}$ has a rational value.

10. Prove Proposition 1.1.2.

11. Let the sides of $\triangle A_0 B_0 C_0$ have lengths 4, 5, 6, and for each $n = 1, 2, 3, \ldots$ let $A_n B_n C_n$ be the triangle formed by joining the midpoints of the sides of $\triangle A_{n-1} B_{n-1} C_{n-1}$. Evaluate the sum of the perimeters of all the triangles $A_n B_n C_n$ for $n = 0, 1, 2, \ldots$

12. Let $A_0 B_0 C_0 D_0$ be a square with side a, and for each $n = 1, 2, 3, \ldots$ let $A_n B_n C_n D_n$ be the square formed by joining the midpoints of the sides of

$A_{n-1}B_{n-1}C_{n-1}D_{n-1}$ in succession. Evaluate the sum of the perimeters of all the squares $A_n B_n C_n D_n$ for $n = 0, 1, 2, \ldots$.

13. Describe the main mathematical achievements of
   a) Archimedes
   b) Apollonius of Perga
   c) Euclid
   d) Eudoxus
   e) Menaechmus

1.2*   THE GEOMETRY OF THE PARABOLA

A modern style argument will now be produced to verify the sequence of propositions which Archimedes used to justify Proposition 1.1.1 and which he did not bother to prove (see Appendix A). The presumption is that his audience was familiar with the missing proofs.

Let $y = ax^2$ be a fixed parabola and let m be a fixed real number. Since the straight lines $y = mx + c$ all have the same slope they are all parallel to each other. Depending on the value of c, these intersect the given parabola in 2, 1, or 0 points, which points are given by the simultaneous solution of the equations

$$y = ax^2$$
$$y = mx + c.$$

The elimination of y in these equations results in the quadratic equation $ax^2 - mx - c = 0$ whose solutions are

$$x_i = \frac{m \pm \sqrt{m^2 + 4ac}}{2a} \qquad i = 1, 2 \tag{3}$$

with corresponding ordinates

$$y_i = mx_i + c \qquad i = 1, 2.$$

Of all these parallel lines only one is tangent to the parabola, namely, that line which intersects it at exactly one point. This point of tangency P is therefore characterized by the equation $x_1 = x_2$ which implies that the radical of (3) is zero. Hence the coordinates of the vertex P are

$$\left(\frac{m}{2a}, a \cdot \left(\frac{m}{2a} \right)^2 \right) = \left(\frac{m}{2a}, \frac{m^2}{4a} \right). \tag{4}$$

Every parabola has a symmetry axis that passes through its vertex. Those straight lines that are parallel to the axis of the parabola are its *diameters* (they bear very little resemblance to the diameters of a circle). The axis of the parabola $y = ax^2$ is the y-axis, and its diameters are the straight lines that are perpendicular to the x-axis. Several interesting conclusions can now be drawn from the above computations.

Proposition 1.2.1 *If QQ' is the base of a segment of a parabola then the diameter through the midpoint of QQ' intersects the segment's arc in its vertex.*

Proof. Let the equation of the chord QQ' be $y = mx + c$. By (3) the midpoint V of QQ' (Figure 1.2) has abscissa

$$\frac{1}{2}(x_1 + x_2) = \frac{1}{2}\left(\frac{m + \sqrt{m^2 + 4ac}}{2a} + \frac{m - \sqrt{m^2 + 4ac}}{2a}\right)$$
$$= \frac{m}{2a}.$$

As noted in Equation (4) this is also the abscissa of the vertex P. Thus, PV, being a vertical line, is a diameter of the parabola. □

Proposition 1.2.2 *The midpoints of a family of parallel chords constitute a diameter of the parabola.*

Proof. This follows immediately from the previous proposition, since all these midpoints must lie directly above the common vertex P. □

The next proposition Archimedes named *the property of the parabola*. In the special case where the base QQ' of the parabolic segment is parallel to the x-axis (see Fig. 1.6), this property reduces to the statement that the ratio

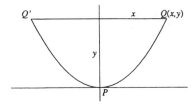

Fig. 1.6 A special parabolic segment

x^2/y is constant, a property that does indeed define the parabola.

Proposition 1.2.3 *Let P be any fixed point of the parabola, let V be any point on the diameter though P, and let QQ' be the chord through V that is parallel to the tangent at P. Then the value of the ratio*

$$\frac{QV^2}{PV}$$

is independent of the position of the point V on the diameter.

Proof. Let the equation of the line QQ' in Figure 1.2 be $y = mx + c$, and let the coordinates of Q and Q' be (x_1, y_1) and (x_2, y_2) respectively. Since V is the midpoint of QQ', $QV^2 = QQ'^2/4$ and it follows from Equation (3) that

$$QQ'^2 = (x_1 - x_2)^2 + (y_1 - y_2)^2 = (x_1 - x_2)^2 + (m(x_1 - x_2))^2$$

$$= (x_1 - x_2)^2(1 + m^2) = \left(2\frac{\sqrt{m^2 + 4ac}}{2a}\right)^2 (1 + m^2)$$

$$= \frac{(m^2 + 4ac)(m^2 + 1)}{a^2}.$$

Since V is the midpoint of QQ' it has ordinate $(y_1 + y_2)/2$ and, as noted in Equation (4) the ordinate of P is $m^2/4a$. Hence,

$$PV = \frac{y_1 + y_2}{2} - \frac{m^2}{4a} = \frac{m(x_1 + x_2) + 2c}{2} - \frac{m^2}{4a}$$

$$= m \cdot \frac{m}{2a} + c - \frac{m^2}{4a} = \frac{m^2 + 4ac}{4a}.$$

It follows that

$$\frac{QV^2}{PV} = \frac{QQ'^2}{4PV} = \frac{1}{4} \cdot \frac{\dfrac{(m^2 + 4ac)(m^2 + 1)}{a^2}}{\dfrac{m^2 + 4ac}{4a}} = \frac{m^2 + 1}{a}.$$

Since $(m^2 + 1)/a$ does not contain the quantity c, it follows that the value of the ratio in question is independent of the position of V on the diameter. $\square$

The next proposition was actually proved by Archimedes. The relevant construction is displayed in Figure 1.7.

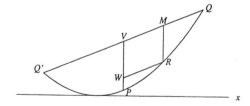

Fig. 1.7 Two succesive bisections of a chord

Proposition 1.2.4 *Let QQ' be a chord of a parabola, let V be the its midpoint, and let M be the midpoint of QV. If the diameters through V and M intersect the parabola in P and R respectively, then*

$$PV = \frac{4}{3}RM.$$

Proof. Draw RW parallel to QQ'. It follows from Proposition 1.2.3 that

$$\frac{QV^2}{PV} = \frac{RW^2}{PW}.$$

Moreover, since $RWVM$ is a parallelogram,

$$QV = 2MV = 2RW.$$

Hence $QV^2 = 4RW^2$ from which it follows that $PV = 4PW$ and so

$$PV = \frac{4}{3}WV = \frac{4}{3}RM.$$

$\square$

Finally, Archimedes set the stage for the limiting process described in the previous section.

Proposition 1.2.5 *Let QQ' be the base of a parabolic segment with vertex P. If R is the vertex of the parabolic segment with base PQ, then*

$$\Delta PQQ' = 8\Delta PRQ.$$

Proof. By Proposition 1.2.2, the diameter through R bisects the chord PQ, say at Y (Fig. 1.8). Thus, it bisects one side PQ of ΔPQV and is parallel to

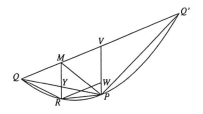

Fig. 1.8 Triangles in parabolic segments

its second side PV (which is also a diameter). Since ΔQYM and ΔQPV are similar, it follows that RY also bisects the third side QV, say at M, and that $YM = (1/2)PV = (1/2) \cdot (4/3)RM = (2/3)RM$. Hence, $YM = 2RY$ and so

$$\Delta PQM = \Delta PYM + \Delta QYM = 2\Delta PYR + 2\Delta QYR = 2\Delta PQR$$
$$\therefore \quad \Delta PQQ' = 2\Delta PQV = 4\Delta PQM = 8\Delta PQR.$$

$\square$

Exercises 1.2

1. A chord of the ellipse $x^2/a^2 + y^2/b^2 = 1$ is a line segment joining any two of its points, and a diameter is a chord that contains the ellipse's center—the

origin. Prove that the midpoints of a family of chords that are all parallel to each other constitute a diameter of the ellipse.

2. A chord of the hyperbola $x^2/a^2 - y^2/b^2 = 1$ is a line segment joining any two of its points, and a diameter is a chord that contains the hyperbola's center—the origin. Prove that the midpoints of a family of chords that are all parallel to each other all lie on a (infinitely extended) diameter of the hyperbola.

3. Prove that of all the points on the arc of a parabola, its vertex has the maximum distance from the base of the corresponding chord. (Actually, Archimedes takes this to be the defining property of the vertex and then quotes a proposition to the effect that the tangent at this point is parallel to the chord).

4. Let X be any point on the parabolic arc $\overset{\frown}{AB}$. Prove that $\triangle ABX \leq 3/4$ of the parabolic segment of $\overset{\frown}{AB}$.

Chapter Summary

In the third century B.C., 1900 years before the advent of modern calculus, Archimedes evaluated the area of the parabolic segment. To accomplish this he used some geometrical propositions about the parabola, a geometrical limiting process, and the infinite geometric progression. This constitutes the first recorded instance of the use of infinite series in general and the infinite geometric progression in particular.

2

Fermat, Differentiation, and Integration

Newton is reported to have said that if he saw farther than other people it was because he stood on the shoulders of giants. If so, then Fermat is to be counted amongst these giants, and his differentiation and integration methods are precursors of the calculus of Leibniz and Newton.

2.1 FERMAT'S CALCULUS

The lawyer and part time mathematician Pierre de Fermat (1601?–1665) is justly famed for his pioneering and influential work in number theory. His contributions to the evolution of calculus are less well known, yet important. His invention of the coordinate system, now known as *Cartesian* coordinates, predates René Descartes' (1596–1650) work on the same topic by eight years. Fermat possessed methods for determining tangents to curves and evaluating the areas of curvilinear regions several years before the births of either Isaac Newton (1642–1727) or Gottfried Wilhelm Leibniz (1646–1716). Yet, Fermat was not alone in anticipating the calculus of Newton and Leibniz. Bonaventura Cavalieri (1598–1647), René Descartes, Blaise Pascal (1623–1662), Evangelista Torricelli (1608–1647), Gilles Persone de Roberval (1602–1675) all had methods for constructing tangents and evaluating areas enclosed by curved lines. This text focuses on Fermat because the well known, and occasionally frustrating, laconic character of his writing makes the exposition of his mathematical work easier.

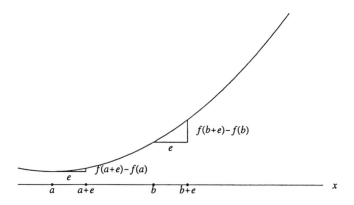

Fig. 2.1   Locating a max/min point

Fermat's method for finding the *maxima* and *minima* of a function $y = f(x)$ was known to him before 1637. His approach was based on the observation that near such a point any small change in the value of x results in a much smaller, and hence "negligible" change in the value of $f(x)$. Thus, in Figure 2.1, both of the depicted triangles have bases of the same length

$$e = (a + e) - a = (b + e) - b.$$

However, the vertical sides, denoting the changes in the value of $f(x)$, are quite different. The quantity $f(a+e) - f(a)$ in the triangle on the left is much smaller than the corresponding quantity $f(b+e) - f(b)$ in the other triangle. This is, of course, a consequence of the fact that at critical points the tangent to the graph is horizontal. The difference between the two vertical sides is in fact qualitative and can be used to locate the critical point. Consider, for example, the function

$$f(x) = x^3 - 2x^2 \tag{1}$$

Let $x + e$ denote a value of the independent variable that is close to x. Fermat's dictum that the difference between $f(x + e)$ and $f(x)$ is negligible translates to the equation

$$f(x + e) - f(x) = 0 \tag{2}$$

or

$$(x + e)^3 - 2(x + e)^2 - x^3 + 2x^2 = 0.$$

This simplifies to

$$3x^2 e + 3xe^2 + e^3 - 4xe - 2e^2 = 0 \tag{3}$$

or, upon division by e,

$$3x^2 + 3xe + e^2 - 4x - 2e = 0. \tag{4}$$

Fermat's method then called for setting $e = 0$. When the resulting equation

$$3x^2 - 4x = 0 \tag{5}$$

was solved, the critical values

$$x = 0 \quad \text{and} \quad x = \frac{4}{3} \tag{6}$$

were obtained.

In this process, Fermat engaged in several practices that violate modern mathematical standards. To begin with, the so called Equation (2) is in reality only an approximation, whereas the solutions of Equation (6) are purported to be exact. This is a sloppiness for which any calculus student would quite properly lose some credit. In addition, when Equation (3) is divided by e, it is tacitly assumed that $e \neq 0$. After all, it is well known that division by zero is an invalid mathematical operation that can lead to absurd results (see Exercise 6). Nevertheless, subsequent to this division, Fermat's method calls for setting $e = 0$ in the resulting Equation (4). This, needless to say, is both inconsistent and incorrect. The same quantity e cannot be both different from and equal to 0, even at different locations on the page. Still, the answers derived in Equation (3) are correct. Equation (5) is, of course, identical with that obtained by equating the derivative of $f(x) = x^3 - 2x^2$ to zero. As this observation indicates, Fermat did indeed possess a method for constructing tangents to arbitrary curves. In other words, he knew how to differentiate.

The crucial observation that near a critical value of x small changes in the value of x yield negligible changes in the values of $f(x)$ goes back at least as far as Johannes Kepler (1571–1630). Other mathematicians of the first half of the sixteenth century were also in possession of methods for locating critical points that suffered from deficiencies similar to those above. It is the concept of the *limit* that is implicit in the manipulations of Fermat and his contemporaries. This notion was made explicit by both Newton and Leibniz and will be discussed in greater detail in Chapter 7.

The lack of rigor in Fermat's argument may come as a surprise to those who have come to expect complete rigor from all mathematical arguments. The fact is many good mathematicians are impatient with the restraints of perfect logic and will often make intuitive, logically questionable, leaps. Fermat was notorious for making such leaps. Posterity justified most of his incompletely verified assertions, though sometimes it took hundreds of years and the cumulative efforts of many mathematicians to do so. In one instance, though, Fermat was demonstrably wrong. He stated that the numbers

$$F_n = 2^{2^n} + 1$$

are prime for $n = 0, 1, 2, 3, \ldots$. This is correct for $F_0 = 3$, $F_1 = 5$, $F_2 = 17$, $F_3 = 257$, $F_4 = 65{,}537$. However, about one hundred years later Leonhard Euler (1707–1783) pointed out that $F_5 = 2^{32} + 1 = 4{,}294{,}967{,}297 = 641 \cdot$

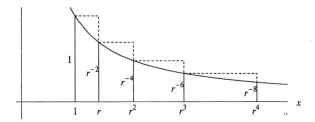

Fig. 2.2 Approximating a curve with rectangles

6,700,417, thus disproving Fermat's assertion. Since then it has been verified that F_n is in fact composite for $n = 5, 6, \ldots, 19$ and for at least 69 other values of n, the largest being $n = 23{,}471$. No further values of n for which F_n is prime have been found.

Fermat used two methods to evaluate of the area under the graph of $y = x^n$. In order to simplify the argument, attention is restricted to the case $n = -2$, leaving the other cases for Exercise 2. Let r be any real number greater than 1, and for each $k = 1, 2, 3, \ldots$, draw the ordinate $(r^k)^{-2} = r^{-2k}$ above the point $x = r^k$ on the x-axis (Fig. 2.2). These ordinates divide the infinite region under the curve and to the right of $x = 1$ into portions that are nearly rectangular and that are completed to exact rectangles by means of the dashed lines in the figure. The first rectangle has area $(r - 1) \times 1$, the second has area $(r^2 - r)r^{-2}$, the third has area $(r^3 - r^2)r^{-4}$ and in general, the kth rectangle has area

$$(r^k - r^{k-1})r^{-2(k-1)} = \frac{r-1}{r^{k-1}}.$$

The sum of the rectangles is now evaluated by means of the infinite geometric progression as

$$(r - 1) + (r - 1)\frac{1}{r} + (r - 1)\left(\frac{1}{r}\right)^2 + (r - 1)\left(\frac{1}{r}\right)^3 + \cdots$$

$$= (r - 1)\left(1 + \frac{1}{r} + \frac{1}{r^2} + \frac{1}{r^3} + \cdots\right)$$

$$= (r - 1)\frac{1}{1 - \dfrac{1}{r}}$$

$$= (r - 1)\frac{r}{r - 1} = r.$$

Consequently, as r approaches 1, the sum of the rectangular areas also approaches 1. Moreover, as r approaches 1, the subdivision points $1, r, r^2, r^3$, $\ldots$, are more and more densely distributed on the x-axis, so that the union of the rectangular regions approaches the region between $y = x^{-2}$ and the x-axis. We now conclude with Fermat that this region has area 1.

Actually, the argument as presented here really only proves that the region in question has area at most 1. To prove equality it is necessary to approximate the curve by a sequence of rectangles below it (see Exercise 5). Fermat was fully aware of this need and stated that the details could be carried out "in the manner of Archimedes."

Exercises 2.1

1. Use Fermat's method to find the critical values of the following functions:

a) $y = x^2 + x$

b) $y = 2x^2 - 3x + 1$

c) $y = x^3 - 3x$

d) $y = x^3 + 3x^2$

e) $y = 2x^3 + 6x^2 - 12$

f) $y = x^3 + 2x^2 + x + 1$

2. Adapt Fermat's integration technique to the evaluation of the following integrals:

a) $\displaystyle\int_1^\infty x^{-n}dx, \quad n > 2$

b) $\displaystyle\int_0^1 x^2 dx \qquad$ (Hint: use $r < 1$)

c) $\displaystyle\int_0^1 x^n dx, \quad n \geq 0$

d) $\displaystyle\int_a^\infty x^{-2}dx, \quad a \geq 0$

e) $\displaystyle\int_a^\infty x^{-n}dx, \quad n > 2, a \geq 0$

f) $\displaystyle\int_0^a x^n dx, \quad n > 0$

3. Apply Fermat's integration procedure (in reverse) to the function e^x to conclude that if $r < 1$, then

$$\lim_{r \to 1}(1 - r)\left(e + re^r + r^2 e^{r^2} + r^3 e^{r^3} + \ldots\right) = e - 1.$$

4. Apply Fermat's integration procedure (in reverse) to the function e^{-x} to conclude that if $r > 1$, then

$$\lim_{r \to 1} (r - 1) \left(e^{-1} + re^{-r} + r^2 e^{-r^2} + r^3 e^{-r^3} + \ldots \right) = e^{-1}.$$

5. Complete Fermat's integration by finding an infinite set of rectangles under the graph of $y = x^{-2}$ in Figure 2.2 the sum of whose areas also converges to 1.

6. Explain the following paradox:

$$3 - 3 = 4 - 4$$
$$\therefore \quad 3(1 - 1) = 4(1 - 1)$$
$$\therefore \quad 3 = 4.$$

7. Describe the main mathematical achievements of
   a) Bonaventura Cavalieri
   b) René Descartes
   c) Pierre de Fermat
   d) Johannes Kepler
   e) Gottfried Wilhelm Leibniz
   f) Isaac Newton
   g) Blaise Pascal
   h) Gilles Personne de Roberval
   i) Evangelista Torricelli

Chapter Summary

Several years before the births of both Newton and Leibniz, Fermat employed procedures that are tantamount to the differentiation and integration of some standard functions. These methods, which were explained only by means of examples, were effective but lacked both generality and rigorous proofs. Fermat's integration of $x^{\pm n}$ again made use of the infinite geometric progression, although in a manner different from that employed by Archimedes.

3

Newton's Calculus
(Part 1)

Modern calculus is generally attributed to Sir Isaac Newton and to Gottfried Wilhelm Leibniz. Though they made their discoveries independently, there was a considerable amount of overlap in their work. This text will present Newton's work in some detail because in the long run his contributions were more influential.

3.1 THE FRACTIONAL BINOMIAL THEOREM

Several major mathematical advances were motivated by investigations relating to the *Binomial Theorem*: Newton's discovery of the power series is a case in point. The modern version of the Binomial Theorem states that if a and b are real numbers and n is a positive integer, then

$$
(a+b)^n = \binom{n}{0}a^n + \binom{n}{1}a^{n-1}b + \binom{n}{2}a^{n-2}b^2 + \dots
$$
$$
+ \binom{n}{k}a^{n-k}b^k + \dots + \binom{n}{n}b^n \tag{1}
$$

where the coefficients $\binom{n}{k}$ are to be computed either directly by means of the formula

$$
\binom{n}{k} = \frac{n(n-1)(n-2)\cdots(n-k+1)}{k(k-1)(k-2)\cdots 1} \tag{2}
$$

or recursively by means of the equations

$$\binom{n}{0} = \binom{n}{1} = 1 \qquad \binom{n}{k} = \binom{n-1}{k} + \binom{n-1}{k-1} \qquad \text{for } n > k > 0. \qquad (3)$$

Equation (3) is the principle that underlies the array commonly known as *Pascal's Triangle*. Exercises 6 and 7 outline a proof of the Binomial Theorem.

As Newton explained (see Appendix D), he was lead to a fractional version of the Binomial Theorem in deriving the area under the curves of the form

$$y = (1 \pm x^2)^{m/n}.$$

He was aware that for integral values of the exponent, i.e., when $n = 1$, the Binomial Theorem yielded

$$\frac{m-0}{1} \cdot \frac{m-1}{2} \cdot \frac{m-2}{3} \cdot \frac{m-3}{4} \cdots \frac{m-(k-1)}{k} \qquad (4)$$

as the coefficient of x^{2k} in the expansion of $(1+x^2)^m$. He also noted that each of these coefficients could be obtained from the previous one by multiplying it with an appropriate fraction. Thus,

$$\frac{m-0}{1} \cdot \frac{m-1}{2} \cdot \frac{m-2}{3} \cdot \frac{m-3}{4} = \left(\frac{m-0}{1} \cdot \frac{m-1}{2} \cdot \frac{m-2}{3} \right) \cdot \frac{m-3}{4}$$

and, in general,

$$\binom{m}{k+1} = \binom{m}{k} \cdot \frac{m-k}{k+1} \qquad (5)$$

(see Exercise 3). By analogy, Newton then reasoned that this recursion would also remain valid for fractional values of m. In other words, he guessed that the first coefficient is 1 and that the subsequent coefficients in the expansion of $(1 \pm x^2)^{m/n}$ satisfy the relation

$$\binom{m/n}{k+1} = \binom{m/n}{k} \cdot \frac{m/n - k}{k+1}.$$

Accordingly, the first few coefficients of $(1 + x^2)^{1/2}$ are

$$\binom{1/2}{0} = 1$$

$$\binom{1/2}{1} = \binom{1/2}{0} \cdot \frac{1/2 - 0}{1} = 1 \cdot \frac{1}{2} = \frac{1}{2}$$

$$\binom{1/2}{2} = \binom{1/2}{1} \cdot \frac{1/2 - 1}{2} = \frac{1}{2} \cdot \left(-\frac{1}{4} \right) = -\frac{1}{8}$$

$$\binom{1/2}{3} = \binom{1/2}{2} \cdot \frac{1/2 - 2}{3} = \left(-\frac{1}{8} \right) \cdot \left(-\frac{1}{2} \right) = \frac{1}{16}$$

$$\binom{1/2}{4} = \binom{1/2}{3} \cdot \frac{1/2 - 3}{4} = \frac{1}{16} \cdot \left(-\frac{5}{8}\right) = -\frac{5}{128}.$$

In short,

$$(1 + x^2)^{1/2} = 1 + \frac{1}{2}x^2 - \frac{1}{8}x^4 + \frac{1}{16}x^6 - \frac{5}{128}x^8 \cdots \tag{6}$$

The reason Newton examined $(1+x^2)^{1/2}$ rather than the simpler expression $(1+x)^{1/2}$ was that the latter was too simple for his purposes. He was looking to integrate new functions and the substitution $u = 1 + x$ converts $(1 + x)^{1/2}$ to $u^{1/2}$ which is easily integrated by means of the formula

$$\int u^n \, du = \frac{u^{n+1}}{n+1} + C$$

which, as will be seen below, Newton already knew.

The extension of the Binomial Theorem to fractional exponents has many uses besides facilitating integration, and so the expansion of $(1 + x)^{1/2}$ is of interest for other reasons. When x^2 is replaced by x, Equation (6) becomes

$$(1 + x)^{1/2} = 1 + \frac{1}{2}x - \frac{1}{8}x^2 + \frac{1}{16}x^3 - \frac{5}{128}x^4 \cdots \tag{7}$$

The replacement of x by $-x^2$ transforms Equation (7) to

$$(1 - x^2)^{1/2} = 1 + \frac{1}{2}(-x^2) - \frac{1}{8}(-x^2)^2 + \frac{1}{16}(-x^2)^3 - \frac{5}{128}(-x^2)^4 \cdots$$

$$= 1 - \frac{1}{2}x^2 - \frac{1}{8}x^4 - \frac{1}{16}x^6 - \frac{5}{128}x^8 \cdots \tag{8}$$

which is one of the expansions that Newton wrote out explicitly (see Appendix D). Newton's fractional version of the Binomial Theorem is stated in full detail. The complete proof appears in Section 14.2.

Theorem 3.1.1 (The Fractional Binomial Theorem) *For every rational number r and non-negative integer k, let the symbol $\binom{r}{k}$ be defined recursively by*

$$\binom{r}{0} = 1, \qquad \binom{r}{k+1} = \frac{r-k}{k+1} \cdot \binom{r}{k} \qquad k = 0, 1, 2, \dots ,$$

then

$$(1 + x)^r = 1 + \binom{r}{1}x + \binom{r}{2}x^2 + \binom{r}{3}x^3 + \dots \qquad \text{for } |x| < 1.$$

It is clear from the above definition that if r is a positive integer, then $\binom{r}{r+1} = 0$ and hence $\binom{r}{k} = 0$ for all $k > r$, so that the binomial expansion is in fact finite. However, for fractional values of r, such as the above $m = 1/2$ (Equation 4), zero will not be encountered, and the expansion will continue ad infinitum. Newton was fully aware that his extension of the Binomial Theorem from integral to fractional values of m was lacking in logical grounding and so he verified Equation (8) by multiplying

$$1 - \frac{1}{2}x^2 - \frac{1}{8}x^4 - \frac{1}{16}x^6 - \frac{5}{128}x^8 \ldots$$

by itself and observing that the result was indeed $1 - x^2$. Note

$$
\begin{array}{ccccc}
1 & -\dfrac{1}{2}x^2 & -\dfrac{1}{8}x^4 & -\dfrac{1}{16}x^6 & -\dfrac{5}{128}x^8 & \cdots \\[2mm]
\times \\[2mm]
1 & -\dfrac{1}{2}x^2 & -\dfrac{1}{8}x^4 & -\dfrac{1}{16}x^6 & -\dfrac{5}{128}x^8 & \cdots
\end{array}
$$

$$
\begin{array}{ccccc}
1 & -\dfrac{1}{2}x^2 & -\dfrac{1}{8}x^4 & -\dfrac{1}{16}x^6 & -\dfrac{5}{128}x^8 & \cdots \\[2mm]
 & -\dfrac{1}{2}x^2 & +\dfrac{1}{4}x^4 & +\dfrac{1}{16}x^6 & +\dfrac{1}{32}x^8 & \cdots \\[2mm]
 & & -\dfrac{1}{8}x^4 & +\dfrac{1}{16}x^6 & +\dfrac{1}{64}x^8 & \cdots \\[2mm]
 & & & -\dfrac{1}{16}x^6 & +\dfrac{1}{32}x^8 & \cdots \\[2mm]
 & & & & -\dfrac{5}{128}x^8 & \cdots
\end{array}
$$

$$1 \quad -x^2.$$

Similarly, having derived the expression

$$(1 - x^2)^{1/3} = 1 - \frac{1}{3}x^2 - \frac{1}{9}x^4 - \frac{5}{81}x^6 \cdots ,$$

Newton multiplied it twice by itself to obtain the value $1 - x^2$.

The extended version of the Binomial Theorem turned out to be valid for negative exponents as well. Thus, according to this theorem,

$$(1+x)^{-1} = 1 + \frac{-1}{1}x + \frac{(-1)(-2)}{1 \cdot 2}x^2 + \frac{(-1)(-2)(-3)}{1 \cdot 2 \cdot 3}x^3$$
$$+ \frac{(-1)(-2)(-3)(-4)}{1 \cdot 2 \cdot 3 \cdot 4}x^4 + \cdots$$
$$= 1 - x + x^2 - x^3 + x^4 + \cdots .$$

Replacing x by $-x$ in this equation results in the well known (and tested) infinite geometric progression:

$$\frac{1}{1-x} = 1 + x + x^2 + x^3 + x^4 + \cdots .$$

Exercises 3.1

1. Find the first 5 terms of the expansion of

a) $(1+x)^{1/3}$ b) $(1-x)^{1/3}$ c) $(1+x^2)^{1/3}$

d) $(1-2x^3)^{1/3}$ e) $(1+x)^{2/3}$ f) $(1-x)^{2/3}$

g) $(1+x^2)^{2/3}$ h) $(1-2x^3)^{2/3}$ i) $(1+x)^{5/3}$

j) $(1-x)^{5/3}$ k) $(1+x^2)^{5/3}$ l) $(1-2x^3)^{5/3}$

m) $(1+x)^{-2}$ n) $(1-x)^{-2}$ o) $(1+x^2)^{-2}$

p) $(1-2x^3)^{-2}$ q) $(1+x)^{-4/3}$ r) $(1-x)^{-4/3}$

s) $(1+x^2)^{-4/3}$ t) $(1-2x^3)^{-4/3}$ u) $(1-x)^{-3}$.

2. Let the binomial coefficient $\binom{n}{k}$ be defined by means of Equation (2). Prove that it satisfies the recurrence of Equation (3).

3. Let the binomial coefficient $\binom{n}{k}$ be defined by means of Equation (2). Prove that it satisfies the recurrence of Equation (5).

4. Explain the relevance of the sequence 11^0, 11^1, 11^2, 11^3, 11^4, to the Binomial Theorem (see Appendix D).

5. Explain the statement of the Binomial Theorem in Appendix C.

6. Prove that if n is a positive integer then $(1+x)^n = \binom{n}{0} + \binom{n}{1}x + \binom{n}{2}x^2 + \cdots + \binom{n}{n}x^n$. (Hint: It is clear that $(1+x)^n = (1+x)(1+x)\cdots(1+x) = d_0 + d_1 x + d_2 x^2 + \cdots + d_n x^n$ for some integers $d_0, d_1, d_2, \cdots, d_n$. That $d_k = \binom{n}{k}$ follows by differentiating this equation k times and setting $x = 0$.)

7. Use Exercise 6 to prove the Binomial Theorem (Equation 1).

3.2   AREAS AND INFINITE SERIES

Newton developed his version of the calculus in the years 1665–1666. Two or three years later he summarized this work in a tract *On Analysis by Infinite Equations (De Analysi...)* that he sent to Isaac Barrow (1630–1677), the Lucasian mathematics professor at Cambridge University. Translated excerpts from this tract appear in Appendix E. In this summary Newton stated his general principles as rules which are quoted and discussed here. The first rule refers to Figure 3.1 where A denotes the origin, $AB = x$, and $BD = y$.

Rule I. *If $ax^{m/n} = y$; it shall be*

$$\frac{an}{m+n}x^{(m+n)/n} = \text{area } ABD.$$

Fig. 3.1 A typical curve.

As Newton's comments towards the end of the tract indicate, he was fully aware of the relationship between the area under a curve and the antiderivative and of the difference between them. This statement about areas should be interpreted as one about antiderivatives, or indefinite integrals, whose modern equivalent is

$$\int x^r dx = \frac{1}{r+1} x^{r+1} + C$$

where r is any real number different from -1. It is clear from Newton's examples that he knew exactly how to handle situations where the curve dips below the x axis, or where A is not the origin.

Rule II. *If the Value of y be made up of several such Terms, the Area likewise shall be made up of the Areas which result from every one of the Terms.*

This is the linearity property of integrals which is written today as

$$\int (af(x) + bg(x))dx = a \int f(x)dx + b \int g(x)dx,$$

where a and b are any two real numbers. These two rules were applied (see Appendix E) in a series of carefully chosen examples to the evaluation of proper and improper definite integrals. Two of these are

$$\int_x^\infty (t^{-2} + t^{-3/2})dt \quad \text{and} \quad \int_1^x \left(2t^3 - 3t^5 - \frac{2}{3}t^{-4} + t^{-3/5} \right) dt.$$

The examples for the third of Newton's rules constitute the main portion of the tract.

Rule III. *But if the value of y, or any of it's Terms be more compounded than the foregoing, it must be reduced into more simple Terms; by performing the Operation in Letters, after the same Manner as Arithmeticians divide in Decimal Numbers, extract the Square Root, or resolve affected Equations...*

For pedagogical reasons the discussion of the first of Newton's examples for Rule III is preceded by the consideration of the special case

$$y = \frac{1}{1+x}. \tag{9}$$

Having noted that Rules I and II are of no avail in finding the area under this curve, we turn to Newton's suggestion for ... *performing the Operation in letters after the same Manner as Arithmeticians....* In particular, Rule III suggests that the same process of long division that is used to get rid of the troublesome denominators of fractions and converts

$$\frac{2345}{31} \quad \text{to} \quad 75.64516129\ldots$$

should be used to convert the function in Equation (9) to a more manageable form. Accordingly, to obtain the quotient when dividing the numerator of 1 by the denominator of $1 + x$, focus on the leading terms of both expressions and obtain a quotient of $1 \div 1 = 1$. The process of long division then yields

$$
1+x \overline{\smash{\big)}\ 1} \\
\underline{1+x} \\
-x + 0.
$$

To divide $1 + x$ into $-x + 0$, focus on the respective leading terms 1 and $-x$ whose quotient is $-x \div 1 = -x$ and the process of long division now yields

$$
1+x \overline{\smash{\big)}\ 1} \\
\underline{1+x} \\
-x + 0 \\
\underline{-x - x^2} \\
x^2 + 0.
$$

At the next stage, the quotient of the leading terms is $x^2 \div 1 = x^2$ and two more repetitions are now easily seen to yield

$$
1+x \overline{\smash{\big)}\ 1} \\
\underline{1+x} \\
-x + 0 \\
\underline{-x - x^2} \\
x^2 + 0 \\
\underline{x^2 + x^3} \\
-x^3 + 0 \\
\underline{-x^3 - x^4} \\
x^4 + 0 \\
\underline{x^4 + x^5} \\
-x^5 + 0.
$$

The above process is summarized as the equation

$$\frac{1}{1+x} = 1 - x + x^2 - x^3 + x^4 - \cdots. \tag{10}$$

Of course, if x is replaced by $-x$, this equation is transformed into the infinite geometric progression

$$\frac{1}{1-x} = 1 + x + x^2 + x^3 + x^4 + \cdots.$$

that was used by both Archimedes and Fermat in their evaluations of areas and that also occurred as a special case of Newton's Fractional Binomial Theorem.

The first example to Rule III actually given by Newton deals with the hyperbola

$$y = \frac{a^2}{b+x}. \tag{11}$$

The leading term of the quotient, a^2/b, is obtained by dividing the numerator a^2 by the first term b of the denominator. The product of this quotient with the whole divisor is $a^2(b+x)/b = a^2 + a^2x/b$ which is then subtracted from the dividend a^2 yielding a remainder of $-a^2x/b$.

$$
\begin{array}{r}
a^2/b \\
b+x \;\overline{\big|\; a^2 } \\
a^2 + a^2x/b \\
\hline
-a^2x/b + 0.
\end{array}
$$

The dividend's second term $-a^2x/b^2$ is obtained by dividing the remainder $-a^2x/b$ by the same b of the denominator, leading to a new remainder of a^2x^2/b^2 and so on. Thus,

$$
\begin{array}{r}
a^2/b - a^2x/b^2 + a^2x^2/b^3 - a^2x^3/b^4 + \cdots \\
b+x \;\overline{\big|\; a^2 } \\
a^2 + a^2x/b \\
\hline
-a^2x/b + 0 \\
-a^2x/b - a^2x^2/b^2 \\
\hline
a^2x^2/b^2 + 0 \\
a^2x^2/b^2 + a^2x^3/b^3 \\
\hline
-a^2x^3/b^3 + 0 \\
-a^2x^3/b^3 - a^2x^4/b^4 \\
\hline
a^2x^4/b^4 \cdots
\end{array}
$$

and so

$$\frac{a^2}{b+x} = \frac{a^2}{b} - \frac{a^2x}{b^2} + \frac{a^2x^2}{b^3} - \frac{a^2x^3}{b^4} \cdots.$$

Newton then used Rules I and II to conclude that the area under the hyperbola (Equation 11) was given by the expression

$$\frac{a^2x}{b} - \frac{a^2x^2}{2b^2} + \frac{a^2x^3}{3b^3} - \frac{a^2x^4}{4b^4} \cdots.$$

As Newton noted, a similar long division yields the series expansion

$$\frac{1}{1+x^2} = 1 - x^2 + x^4 - x^6 + x^8 \cdots .$$

Actually, the same series may be obtained simply by replacing the x of Equation (10) with x^2. Consequently, it follows from Rules I and II that the area under the graph of $y = (1 + x^2)^{-1}$, from the origin to the point x on the x axis, is

$$x - \frac{1}{3}x^3 + \frac{1}{5}x^5 - \frac{1}{7}x^7 \cdots .$$

However, this same area is recognized as

$$\int \frac{dx}{1+x^2} = \tan^{-1} x + C.$$

Since the area in question is 0 when $x = 0$, it follows that $C = 0$ so that we get

$$\tan^{-1} x = x - \frac{1}{3}x^3 + \frac{1}{5}x^5 - \frac{1}{7}x^7 \cdots , \tag{12}$$

a series that was first derived by James Gregory (1638–1675). If we now set $x = 1$ in Gregory's series and recall that $\tan^{-1} 1 = \pi/4$ we obtain the series of Leibniz

$$\frac{\pi}{4} = 1 - \frac{1}{3} + \frac{1}{5} - \frac{1}{7} \cdots .$$

This last series is intriguing and will be justified in Section 14.1. While it looks as though it could be used for evaluating the decimal expansion of π, this is not the case in practice. Estimates derived by calculating the partial sums of this series are not very accurate. Nevertheless, in combination with some other information, Gregory's series (Equation 12) does indeed provide a practical method for evaluating π with great accuracy (Exercises 2, 3).

Newton took the trouble to point out that in converting $1/(1 + x^2)$ to an infinite series by means of long division, the decision to consider 1 as the leading term of the denominator $1 + x^2$ was subjective. If x^2 were taken as the leading term of this fraction the long division process would yield

$$\frac{1}{1+x^2} = \frac{1}{x^2+1} = x^{-2} - x^{-4} + x^{-6} - x^{-8} \cdots ,$$

an expression, he noted, that is useful when the values of x are large.

Newton also mentioned, without going into details, that the process of long division can be used to convert

$$y = \frac{2x^{1/2} - x^{3/2}}{1 + x^{1/2} - 3x} \tag{13}$$

to

$$2x^{1/2} - 2x + 7x^{3/2} - 13x^2 + 34x^{5/2} \cdots . \tag{14}$$

The following are the details omitted by Newton. The required computations can be simplified by setting $u = x^{1/2}$ so that Equation (13) becomes

$$y = u\frac{2 - u^2}{1 + u - 3u^2}.$$

To apply long division to the fraction $(2-u^2)/(1+u-3u^2)$ focus on the leading terms of both the numerator and the denominator and obtain $2 \div 1 = 2$ as the first term of the quotient. The required long division is

$$
\begin{array}{r|lllllll}
 & 2 & - & 2u & + & 7u^2 & - & 13u^3 & + & 34u^4 & \cdots \\
\hline
1 + u - 3u^2 & 2 & + & 0u & - & u^2 \\
 & 2 & + & 2u & - & 6u^2 \\
\hline
 & & & -2u & + & 5u^2 \\
 & & & -2u & - & 2u^2 & + & 6u^3 \\
\hline
 & & & & & 7u^2 & - & 6u^3 \\
 & & & & & 7u^2 & + & 7u^3 & - & 21u^4 \\
\hline
 & & & & & & & -13u^3 & + & 21u^4 \\
 & & & & & & & -13u^3 & - & 13u^4 & + & 39u^5 \\
\hline
 & & & & & & & & & 34u^5 & - & 39u^5.
\end{array}
$$

Consequently

$$\frac{2 - u^2}{1 + u - 3u^2} = 2 - 2u + 7u^2 - 13u^3 + 34u^4 \cdots$$

from which expression (14) is easily obtained. The same expression could also have been obtained in a different manner. It follows from Equation (10) and several applications of the Binomial Theorem that

$$
\begin{aligned}
\frac{1}{1 + u - 3u^2} \\
= \frac{1}{1 + (u - 3u^2)} \\
= 1 - (u - 3u^2) + (u - 3u^2)^2 - (u - 3u^2)^3 + (u - 3u^2)^4 \cdots \\
= 1 - (u - 3u^2) + (u^2 - 6u^3 + 9u^4) - (u^3 - 9u^4 + \cdots) + (u^4 - \cdots) \cdots \\
= 1 - u + 4u^2 - 7u^3 + 19u^4 \cdots
\end{aligned}
$$

where the omitted portions contain terms of degree 5 or higher. Consequently,

$$
\begin{aligned}
\frac{2 - u^2}{1 + u - 3u^2} &= (2 - u^2)(1 - u + 4u^2 - 7u^3 + 19u^4 \cdots) \\
&= 2 - 2u + 7u^2 - 13u^3 + 34u^4 \cdots .
\end{aligned}
$$

Newton's Rule III also refers to analogs of decimal methods for the extraction of roots and the solution of equations. There was at that time a well known procedure for extracting square (and higher order) roots that was commonly taught in high schools until very recently, and whose details were so onerous as to make the long division process seem like child's play. Nowadays, however, subsequent to the advent of the hand-held calculators, few are familiar with it, and so there is no further discussion beyond its inclusion in Appendix E. On the other hand, the utility of infinite series in the solution of equations, a topic to which Newton dedicates a major portion of the *De Analysi* tract, is the subject matter of the next chapter.

Exercises 3.2

1. Find the first 5 non zero terms of the infinite series expansion of

a) $\dfrac{1}{1-2x}$　　b) $\dfrac{1}{2+x}$　　c) $\dfrac{2}{4x+3}$

d) $\dfrac{1}{1-x-x^2}$　　e) $\dfrac{1}{1-x^2}$　　f) $\dfrac{1}{1+x-2x^2}$

g) $\dfrac{x+1}{1-9x^2}$　　h) $\dfrac{x+1}{x^2+x+1}$　　i) $\dfrac{1+x}{4-x^2}$

j) $\dfrac{1+2x}{1+x^3}$　　k) $\dfrac{1+x+x^2}{1+x^2+x^4}$　　l) $\dfrac{2+3x}{1+x-x^2+x^3}$.

2. Use the trigonometric formula

$$\tan(A+B) = (\tan A + \tan B)/(1 - \tan A \tan B)$$

to prove that $\pi/4 = \tan^{-1}(1/2) + \tan^{-1}(1/3)$. Use Equation (12) through the x^{11} term and a calculator (with 8 decimal places at least) to estimate $4(\tan^{-1}(1/2) + \tan^{-1}(1/3))$. Compare the resulting estimate of π with its actual value.

3. Use the trigonometric formula

$$\tan(A+B) = (\tan A + \tan B)/(1 - \tan A \tan B)$$

to prove that $\pi/4 = 4\tan^{-1}(1/5) - \tan^{-1}(1/239)$. Use Equation (12) through the x^7 term and a calculator (with 8 decimal places at least) to estimate $4(4\tan^{-1}(1/5) - \tan^{-1}(1/239))$. Compare the resulting estimate of π with its actual value.

4. Assuming that $\sin^{-1}x = \int(1-x^2)^{-1/2}dx$, obtain a power series expansion for $\sin^{-1}x$.

5. Assuming that $\ln(1+x) = \int(1+x)^{-1}dx$, obtain a power series expansion for $\ln(1+x)$.

6. Describe the main mathematical achievements of
   a) Isaac Barrow
   b) James Gregory.

3.3   NEWTON'S PROOFS

De Analysi, Newton's first written description of the calculus, contains very little by way of formal proofs. Instead, he relied on lengthy computational verifications of specific examples to check on the validity of those conclusions to which his reasoning by analogy led. As his opening sentence states: "The General Method, which I had devised some considerable Time ago, for measuring the Quantities of Curves, by Means of Series, infinite in the Number of their Terms, is rather shortly explained, than accurately demonstrated in what follows." Nevertheless, in the concluding paragraphs, he did offer some arguments to support his methods. In subsequent works he formulated a fairly rigorous theory of differentiation but he never returned to the issue of the convergence of series.

The following paragraphs are lightly edited extracts from these arguments. A more faithful rendition of the original text appears in Appendix E. Newton's argument for Rule I is quite reminiscent of Fermat's max/min method. However after discussing the specific example of the area under the curve $y = x^{1/2}$, Newton gives a more general argument for $y = x^{m/n}$. Figure 3.2 illustrates this proof.

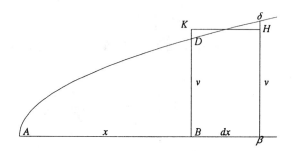

Fig. 3.2   The area as an antiderivative.

Let then $AD\delta$ be any Curve whose Base $AB = x$, the perpendicular Ordinate $BD = y$ and the Area $ABD = z$, as at the Beginning. Likewise put $B\beta = dx$, $BK = v$; and the rectangle $B\beta HK$ $((dx)v) =$ Space $B\beta\delta D$.

Therefore it is $A\beta = x + dx$ and $A\delta\beta = z + (dx)v$. Which things being premised, assume any Relation betwixt x and z that you please and seek for y in the following Manner.

Take, at Pleasure, $\frac{2}{3}x^{3/2} = z$; or $\frac{4}{9}x^3 = z^2$. Then $x + dx(A\beta)$ being substituted for x and $z + (dx)v(A\delta\beta)$ for z, there arises

$$\frac{4}{9}(x^3 + 3x^2 dx + 3x(dx)^2 + (dx)^3) = z^2 + 2zv(dx) + (dx)^2 v^2$$

(from the Nature of the Curve). And taking away Equals $\left(\frac{4}{9}x^3 \text{ and } z^2\right)$ and dividing the Remainders by dx, there arises

$$\frac{4}{9}(3x^2 + 3x(dx) + 3(dx)^2) = 2zv + (dx)v^2.$$

Now if we suppose $B\beta$ to be diminished infinitely and to vanish, or dx to be nothing, v and y in that Case will be equal, and the Terms which are multiplied by dx will vanish: So that there will remain

$$\frac{4}{9} \cdot 3x^2 = 2zv \quad \text{or} \quad \frac{2}{3}x^2 \; (= zy) = \frac{2}{3}x^{3/2}y; \quad \text{or} \quad x^{1/2} \; (= x^2/x^{3/2}) = y.$$

Wherefore conversely, if it be $x^{1/2} = y$, it shall be $\frac{2}{3}x^{3/2} = z$.

Or, universally, if $[n/(m+n)]ax^{(m+n)/n} = z$; or putting $na/(m+n) = c$ and $m + n = p$, if $cx^{p/n} = z$; or $c^n x^p = z^n$: Then by substituting $x + dx$ for x and $z + (dx)v$ (or which is the same $z + (dx)y$) for z, there arises

$$c^n(x^p + p(dx)x^{p-1}\cdots) = z^n + n(dx)yz^{n-1}\cdots,$$

the other Terms, which would at length vanish being neglected. Now taking away $c^n x^p$ and z^n which are equal and dividing the Remainders by dx, there remains

$$c^n p x^{p-1} = nyz^{n-1} \; (= nyz^n/z) = nyc^n x^p/cx^{p/n}.$$

or, by dividing by $c^n x^p$, it shall be

$$px^{-1} = ny/cx^{p/n} \quad \text{or} \quad pcx^{(p-n)/n} = ny;$$

or by restoring $na/(m+n)$ for c and $m+n$ for p, that is m for $p-n$ and na for pc, it becomes $ax^{m/n} = y$. Wherefore conversely, if $ax^{m/n} = y$, it shall be $[n/(m+n)]ax^{(m+n)/n} = z$.   Q.E.D.

The above essentially valid derivation of the standard differentiation formula $(x^r)' = rx^{r-1}$ is followed in the conclusion of *De Analysi* by a considerably less satisfactory discussion of the convergence of power series. Here Newton seems to be trying to mimic the method used by Archimedes to argue that the successive polygons inscribed in the parabolic segment converge to it. This argument, as was mentioned in Section 1.1, was based on the principle that if one repeatedly removes more than half of a quantity the remainder converges to zero. Newton's discussion makes use of the easily verified (see Exercise 1a) fact that for any positive integer n,

$$\left(\frac{1}{2}\right)^n = \frac{1}{2}\left[\left(\frac{1}{2}\right)^n + \left(\frac{1}{2}\right)^{n+1} + \left(\frac{1}{2}\right)^{n+2} + \left(\frac{1}{2}\right)^{n+3} + \cdots\right].$$

Here are Newton's own words:

> Thus, if it be $x = 1/2$, you have x the half of all these $x + x^2 + x^3 + x^4 \cdots$ and x^2 the half of all these $x^2 + x^3 + x^4 + x^5 \cdots$. Therefore if $x < 1/2$

x shall be greater than half of all these $x + x^2 + x^3 \cdots$ and x^2 greater than half of all these $x^2 + x^3 + x^4 \cdots$. Thus if $x/b < 1/2$, you shall have x more than half of all these $x + x^2/b + x^3/b^2 \cdots$, And the same Way of others. And as to the numerical Coefficients, for most part they perpetually decrease: Or if at any Time they increase, you need only suppose x some few Times less.

If the readers feel that this argument is weak they are right. After all, the infinite geometric progression also converges for values of x that are greater than $1/2$, to which values Newton's argument does not apply. Moreover, modern standards of rigor cannot be satisfied with a proof that uses the phrase "...for the most part...". Chapters 9, 10, 13 and 14 will supply the details missing from Newton's tract. In defense of Newton's arguments it should be pointed out that several of these modern proofs regarding the behavior of the general power series are indeed based on comparisons with the infinite geometric progression.

Exercises 3.3

Use the infinite geometrical progression to verify the following assertions made by Newton.

1. For every positive integer n,

a) $\left(\dfrac{1}{2}\right)^n = \dfrac{1}{2}\left[\left(\dfrac{1}{2}\right)^n + \left(\dfrac{1}{2}\right)^{n+1} + \left(\dfrac{1}{2}\right)^{n+2} + \left(\dfrac{1}{2}\right)^{n+3} + \cdots\right].$

b) $x^n > \frac{1}{2}(x^n + x^{n+1} + x^{n+2} + x^{n+3} + \cdots)$ if $0 < x < \dfrac{1}{2}$.

c) $a_n x^n > \frac{1}{2}\left(a_n x^n + a_{n+1} x^{n+1} + a_{n+2} x^{n+2} + a_{n+3} x^{n+3} + \cdots\right)$ if $0 < x < \dfrac{1}{2}$ and $a_1 \geq a_2 \geq a_3 \geq a_4 \geq \cdots \geq 0$.

d) $a_n x^n > \frac{1}{2}\left(a_n x^n + a_{n+1} x^{n+1} + a_{n+2} x^{n+2} + a_{n+3} x^{n+3} + \cdots\right)$ if $0 < x < \dfrac{1}{4}$ and $0 \leq a_{k+1} \leq 2a_k$ for each $k = 1, 2, 3, \ldots$

e) $a_n x^n > \frac{1}{2}\left(a_n x^n + a_{n+1} x^{n+1} + a_{n+2} x^{n+2} + a_{n+3} x^{n+3} + \cdots\right)$ if $b > 0$, $0 < x < \dfrac{1}{2b}$, and $0 \leq a_{k+1} \leq b a_k$ for each $k = 1, 2, 3, \ldots$

Chapter Summary

Isaac Newton is credited with consolidating the disparate methods of several of his predecessors and contemporaries into a unified branch of mathematics that became known as Calculus. This corpus related to both differentiation (and its reverse, integration) and the topic of (infinite) power series. He was led to many of his major contributions to this topic by reasoning by analogy. In particular, he figured that the same binomial theorem that was known to hold for positive integer exponents should also hold for arbitrary rational exponents. Similarly, the four arithmetical operations that hold for real numbers should also hold for variables. Both of these assumptions led him to infinite power series. Newton did offer some arguments to support his new mathematics, but these remain ultimately unsatisfactory.

$$\frac{4}{}$$

Newton's Calculus (Part 2)

As Newton recognized and stressed, power series provide a powerful technique for solving a variety of algebraic and differential equations.

4.1 THE SOLUTION OF DIFFERENTIAL EQUATIONS

An *ordinary differential equation* is an equation that relates x, y, y' and possibly higher derivatives. Such equations are

$$y' = x + y$$
$$y' = 1 - 3x + y + x^2 + xy$$
$$(1 + x^2)y'' + 2xy' + 4x^2y = 0.$$

· Many important mathematical and scientific problems can be reduced to the solution of such equations. Newton knew this and he devoted several pages of *"On the Method of Fluxions..."* to their study. In these he presented a variety of techniques for solving such differential equations. The most general of these calls for replacing y and its derivatives by power series with unknown coefficients and then using the given differential equation to evaluate these coefficients. This is the only one of Newton's techniques that will be discussed in this section by means of several examples that do not involve second or higher order derivatives.

The general outline of his method is the same for all of the examples and seems eminently reasonable. In fact, however, several of the steps need justifications that will only be provided in Chapters 13 and 14. The variable y

in the proposed differential equation is assumed to have an (unknown) power series expansion in terms of x. The power series of y' is then obtained by differentiating this power series term by term and both y and y' are replaced by their series expansions. The given differential equation is then broken down into an infinite number of equations, one for each power x^n and, if possible, these derived equations are used to evaluate the coefficients of the power series of y.

Example 4.1.1 Consider the differential equation $y' = x + y$. Suppose the solution of this equation has the infinite series expansion

$$y = c_0 + c_1 x + c_2 x^2 + c_3 x^3 + c_4 x^4 + \cdots = \sum_{n=0}^{\infty} c_n x^n. \tag{1}$$

Term by term differentiation yields

$$y' = c_1 + 2c_2 x + 3c_3 x^2 + 4c_4 x^3 + \cdots = \sum_{n=1}^{\infty} n c_n x^{n-1}$$

$$= \sum_{n=0}^{\infty} (n+1) c_{n+1} x^n. \tag{2}$$

The substitution of Equations (1) and (2) into the given equation transforms the latter into

$$\sum_{n=0}^{\infty} (n+1) c_{n+1} x^n = x + \sum_{n=0}^{\infty} c_n x^n.$$

With Newton we take it for granted that for each power x^n the corresponding coefficients on the two sides of the equation must balance out. This results in the following infinite number of *level equations:*

Level	Equation	Recurrence
x^0	$c_1 = c_0$	$c_1 = c_0$
x^1	$2c_2 = 1 + c_1$	$c_2 = \dfrac{1 + c_1}{2}$
$x^n \ (n \geq 2)$	$(n+1)c_{n+1} = c_n$	$c_{n+1} = \dfrac{c_n}{n+1}$

Hence, if c_0 is replaced by c, we have

$$c_1 = c,$$

$$c_2 = \frac{1+c}{2},$$

$$c_3 = \frac{1+c}{2 \cdot 3},$$

$$c_4 = \frac{1+c}{2 \cdot 3 \cdot 4},$$

$$c_n = \frac{1+c}{n!} \quad \text{for } n \geq 2.$$

In other words, the general solution to the given differential equation is the function

$$y = c + cx + (1+c)\left(\frac{x^2}{2!} + \frac{x^3}{3!} + \frac{x^4}{4!} + \cdots\right).$$

Example 4.1.2 Consider the differential equation $y' = 1 - 3x + y + x^2 + xy$. Once again it is assumed that y and y' satisfy Equations (1) and (2) respectively, so that the given equation is transformed into

$$\sum_{n=0}^{\infty}(n+1)c_{n+1}x^n = 1 - 3x + x^2 + \sum_{n=0}^{\infty}c_n x^n + x\sum_{n=0}^{\infty}c_n x^n$$

$$= 1 - 3x + x^2 + \sum_{n=0}^{\infty}c_n x^n + \sum_{n=0}^{\infty}c_n x^{n+1}$$

$$= 1 - 3x + x^2 + \sum_{n=0}^{\infty}c_n x^n + \sum_{n=1}^{\infty}c_{n-1}x^n$$

This translates to the table:

Level	Equation	Recurrence
x^0	$c_1 = 1 + c_0$	$c_1 = 1 + c_0$
x^1	$2c_2 = -3 + c_1 + c_0$	$c_2 = \dfrac{-3 + c_1 + c_0}{2}$
x^2	$3c_3 = 1 + c_2 + c_1$	$c_3 = \dfrac{1 + c_2 + c_1}{3}$
x^n $(n \geq 3)$	$(n+1)c_n = c_n + c_{n-1}$	$c_{n+1} = \dfrac{c_n + c_{n-1}}{n+1}$

Hence, if we replace c_0 by c, we get

$$c_1 = 1 + c,$$

$$c_2 = -1 + c,$$

$$c_3 = \frac{1+2c}{3},$$

$$c_4 = \frac{-2+5c}{12},$$

$$c_5 = \frac{2+13c}{60},$$

$$\cdots.$$

The particular case $c = 0$ yields the solution derived by Newton

$$y = x - x^2 + \frac{x^3}{3} - \frac{x^4}{6} + \frac{x^5}{30} \cdots .$$

Example 4.1.3 This section's method can be used to provide Newton's Fractional Binomial Theorem with a somewhat firmer grounding than the reasoning by analogy given in Section 3.1. Observe that the function $y = (1 + x)^r$ has the derivative $y' = r(1+x)^{r-1}$ and hence it is a solution of the differential equation

$$r \cdot y = (1 + x) \cdot y'. \tag{3}$$

The substitution of Equations (1) and (2) into this equation yields

$$r \sum_{n=0}^{\infty} c_n x^n = (1 + x) \sum_{n=1}^{\infty} n c_n x^{n-1} = \sum_{n=1}^{\infty} n c_n x^{n-1} + \sum_{n=1}^{\infty} n c_n x^n$$

$$= \sum_{n=0}^{\infty} (n + 1) c_{n+1} x^n + \sum_{n=0}^{\infty} n c_n x^n.$$

The level k equation is

$$r c_k = (k + 1) c_{k+1} + k c_k, \qquad k = 0, 1, 2, \cdots$$

or

$$c_{k+1} = \frac{r - k}{k + 1} \cdot c_k, \qquad k = 0, 1, 2, \cdots .$$

The choice of the particular value $c_0 = 1$ then yields

$$c_k = \binom{r}{k} \qquad k = 0, 1, 2 \cdots .$$

Thus, both the binomial series and the function $(1 + x)^r$ are solutions of the same differential Equation (3) that have the same initial value of 1 at $x = 0$. It therefore seems reasonable to conclude that they are equal and that we have here a proof of Theorem 3.1.1.

There are several problems with this purported proof of the Fractional Binomial Theorem. In the first place, unless r is a positive integer, the binomial series never converges for $x > 1$ and at this point it is quite conceivable that for some values of r it may never converge at all. Thus, it is not clear that the binomial series is in fact a function in any sense of the word. This problem will be resolved in Example 10.1.5, where it will be demonstrated that, regardless of the value of the exponent r, the binomial series always converges for $-1 < x < 1$.

Another deficiency of this argument is that even after the binomial series is proven to actually define a function, it is not immediately clear that this function is in fact differentiable. That this is indeed the case is proven in Corollary 13.2.10.

Finally, the overall logic of this argument is that once the initial value at $x = 0$ has been specified, there is only one solution to any given first order differential equation. This need not be the case, as is demonstrated by the two functions

$$f(x) = 0 \quad \text{and} \quad f(x) = \frac{x^2}{4}$$

both of which are solutions of the initial value problem

$$y' = y^{1/2}, \qquad y(0) = 0.$$

This last difficulty will be resolved by Corollary 14.2.7.

In conclusion it is necessary to point out that the method presented in this section is not infallible. Exercise 6 exhibits an instance of the failure of this method, or rather, indicates that it sometimes requires further refinements.

Exercises 4.1

1. Use the method of infinite series to solve the following differential equations:

a) $y' = x - y$

b) $y' = 1 + x + y$

c) $y' = 2 - x + xy$

d) $1 + 2x - 2y + 3y' = xy$

e) $y' = 1 + xy$

f) $y' = 2 - x^2 y$

g) $y' = y + \dfrac{1}{1 - x}$

h) $y' = y + \dfrac{x}{1 + x}$

i) $y' = y + \sqrt{1 + x}$

j) $y' = y + \tan^{-1} x$

k) $xy' = 1 + y$

l) $x^2 y' = 1 - y$

m) $3 - x^2 + 2xy = y' + \dfrac{1}{1 - x}$

n) $y' = \dfrac{1}{1 + x^2}$, $y(0) = 0$

o) $y' = \dfrac{1}{\sqrt{1 - x^2}}$ $y(0) = 0$

2. Use the method of infinite series to solve the following differential equations:

a) $y'' - 2xy' + y = 0$

b) $y'' = xy$

c) $y'' + y' + y = x$

d) $y'' + y' + y = \dfrac{1}{1 - x}$

e) $y'' - 3y' + 2y = 0$

f) $y'' - 3y' - 10y = x$

3. Use the fact that $y = \sin x$ satisfies the differential equation $y'' = -y$ to show that

$$\sin x = x - \frac{x^3}{3!} + \frac{x^5}{5!} - \frac{x^7}{7!} + \frac{x^9}{9!} \cdots .$$

4. Use the fact that $y = \cos x$ satisfies the differential equation $y'' = -y$ to show that

$$\cos x = 1 - \frac{x^2}{2!} + \frac{x^4}{4!} - \frac{x^6}{6!} + \frac{x^8}{8!} \cdots .$$

5. Use the fact that $y = e^x$ satisfies the differential equation $y' = y$ to show that

$$e^x = 1 + \frac{x}{1!} + \frac{x^2}{2!} + \frac{x^3}{3!} + \frac{x^4}{4!} \cdots .$$

6. Show that $y = 1/x$ is a solution of the differential equation $xy' + y = 0$ that cannot be obtained directly by this section's method. Explain how the substitution $u = 1 + x$ resolves this difficulty.

4.2*   THE SOLUTIONS OF ALGEBRAIC EQUATIONS

By the time Newton was developing his version of calculus, mathematicians had devised a variety of numerical methods for solving algebraic equations such as

$$x^7 - 5x^4 + 3x^2 - 2 = 0.$$

These methods employed recursive procedures in which each estimate to the solution was used to obtain a better estimate, much like the Newton–Raphson method that is commonly taught in first year calculus. In his tract *De Analysi* Newton explained one such technique that is actually different from the aforementioned Newton–Raphson method. He then went on to show how the same technique could be used to solve an algebraic equation such as

$$y^3 + y - 2 + xy - x^3 = 0 \tag{4}$$

for y. The solution here is no longer a number, of course, but a power series expansion of y in terms of x.

The overall strategy is the same for these algebraic equations as for the differential equations of the previous section. It is first assumed that a power series expansion of the form of Equation (1) actually exists. This series is then substituted into the given equation and, by focusing on the successive powers of x, the coefficients of the power series are evaluated one at a time. Unfortunately, the given equation may include powers of y, leading to powers of Equation (1) and a situation where the resultant level equations are difficult to describe. This complication calls for a further refinement of the method. Rather than write down all of the resultant level equations and then solve for as many coefficients as needed, these level equations are derived, and solved, one at a time. As each of the level equations is solved, the solution is used to modify the given equation and prepare it for the next iteration of the process. The formal description of this process necessitates two definitions. First, assuming Equation (1), we define the *tail series*

$$t_n = c_n x^n + c_{n+1} x^{n+1} + c_{n+2} x^{n+2} + \cdots, \qquad n = 0, 1, 2, \ldots . \qquad (5)$$

Clearly,

$$t_0 = y \quad \text{and} \quad t_n = c_n x^n + t_{n+1}, \qquad \text{for all } n.$$

Second, the given algebraic equation will be assumed to be

$$F(x, y) = 0.$$

As the given equation will be modified in each iteration of the solution process, we begin by setting

$$F_0(x, y) = F(x, y).$$

The *k*th *iteration* of the solution process consists of the execution of the following three steps:

Step 1: Extract the x^k level equation out of $F_k(x, t_k) = 0$;

Step 2: Solve the x^k level equation for c_k;

Step 3: Substitute

$$t_k = c_k x^k + t_{k+1}$$

into $F_k(x, t_k) = 0$ and use the Binomial Theorem to obtain the new equation $F_{k+1}(x, t_{k+1})$.

The logic and limitations of this iterative procedure will be discussed following an example.

Example 4.2.1 Consider the equation

$$y^3 - x - 1 = 0.$$

Here we have

$$F_0(x, t_0): \qquad\qquad\qquad t_0^3 - x - 1 = 0$$

Iteration #0

Step 1: Since $t_0 = c_0 + c_1 x + c_2 x^2 + \cdots$, and since the x^0 level equation ignores all non constant terms in $F_0(x, t_0)$, it follows that

$$x^0 \text{ level:} \qquad\qquad\qquad c_0^3 - 1 = 0$$

Step 2: The only real solution of this equation is $c_0 = 1$.

Step 3: In $F_0(x, t_0)$ above replace t_0 with $1 + t_1$ to obtain

$$(1 + t_1)^3 - x - 1 = 0$$

which simplifies to

$$F_1(x, t_1): \qquad\qquad 3t_1 + 3t_1^2 + t_1^3 - x = 0.$$

Iteration #1

Step 1: Since $t_1 = c_1 x + c_2 x^2 + c_3 x^3 + \cdots$ and since the x^1 level equation ignores all terms of degree >1 in $F_1(x, t_1)$, it follows that

x^1 level: $\qquad\qquad 3c_1 - 1 = 0$

Step 2: $c_1 = \dfrac{1}{3}$

Step 3: In $F_1(x, t_1)$ replace t_1 with $x/3 + t_2$ to obtain

$$3\left(\frac{x}{3} + t_2\right) + 3\left(\frac{x}{3} + t_2\right)^2 + \left(\frac{x}{3} + t_2\right)^3 - x = 0$$

which simplifies to

$$F_2(x, t_2): \qquad 3t_2 + 3t_2^2 + t_2^3 + 2t_2 x + t_2^2 x + \frac{x^2}{3} + \frac{t_2 x^2}{3} + \frac{x^3}{27} = 0.$$

Iteration #2

Step 1: Since $t_2 = c_2 x^2 + c_3 x^3 + c_4 x^4 + \cdots$ and since the x^2 level equation ignores all terms of degree >2 in $F_2(x, t_2)$, it follows that

x^2 level: $\qquad\qquad 3c_2 + \frac{1}{3} = 0$

Step 2: $c_2 = -\dfrac{1}{9}$

Step 3: Stop.

Thus,

$$y = 1 + \frac{x}{3} - \frac{x^2}{9} \cdots$$

is an approximate solution to the given equation.

The fractional Binomial Theorem affords us with another way to solve this example's equation. Accordingly,

$$y = \sqrt[3]{1+x} = (1+x)^{1/3} = 1 + \binom{1/3}{1}x + \binom{1/3}{2}x^2 \cdots$$

$$= 1 + \frac{1/3}{1}x + \frac{1/3}{1} \cdot \frac{1/3 - 1}{2}x^2 \cdots = 1 + \frac{x}{3} - \frac{x^2}{9} \cdots.$$

We now restate and comment on the three steps in each iteration of Newton's procedure.

Step 1: Extract the x^k level equation out of $F_k(x, t_k) = 0$. If the term t_k in the equation $F_k(x, t_k) = 0$ were replaced by its power series expansion of Equation (5), each summand would have x^k as a factor. The reason for this is that the Step 3 substitutions of iterations $0, 1, 2, \ldots, k - 1$ have eliminated all the terms of degrees $0, 1, 2, \ldots, k - 1$. Consequently, the only summands of $F_k(x, t_k)$ that make a contribution to the x^k level equation have forms At_k and Bx^k for some numbers A, B. In other words, the x^k level equation has the form

$$Ac_k + B = 0.$$

Step 2: Solve the x_k level equation for c_k. If the number A above is not zero we get

$$c_k = -\frac{B}{A}.$$

If A is zero, the method fails, and a refinement is needed. This refinement is known as Newton's Polygon Method and falls outside the scope of this text.

Step 3: Substitute

$$t_k = c_k x^k + t_{k+1}$$

into $F_k(x, t_k) = 0$ and use the Binomial Theorem to obtain the new equation $F_{k+1}(x, t_{k+1})$. By the time we get to iteration #3 this step is of course quite messy and it was this complexity that dictated our decision to stop after iteration #2. Computerized symbolic manipulators can be quite helpful here. All of this section's examples are redone and extended at the end with the aid of the application Mathematica.

Example 4.2.2 Consider the equation $y^3 + y - 2 + xy - y^3 = 0$. Replacing y with t_0 yields

$F_0(x, t_0)$: $t_0^3 + t_0 - 2 + t_0 x - x^3 = 0$.

Iteration #0

Step 1: Ignoring all the non constant terms in $F_0(x, t_0)$, we get

x^0 level: $c_0^3 + c_0 - 2 = 0$

Step 2: The only real solution of this equation is $c_0 = 1$.

Step 3: In $F_0(x, t_0)$ above replace t_0 with $1 + t_1$ to obtain

$$(1 + t_1)^3 + (1 + t_1) - 2 + x(1 + t_1) - x^3 = 0$$

which simplifies to

$F_1(x, t_1)$: $4t_1 + 3t_1^2 + t_1^3 + x + t_1 x - x^3 = 0$.

Iteration #1

Step 1: Ignoring all x^n with $n > 1$ in $F_1(x, t_1)$ yields

x^1 level: $4c_1 + 1 = 0$

Step 2: $c_1 = -\dfrac{1}{4}$.

Step 3: In $F_1(x, t_1)$ replace t_1 with $-x/4 + t_2$ to obtain

$$4\left(-\frac{x}{4} + t_2\right) + 3\left(-\frac{x}{4} + t_2\right)^2 + \left(-\frac{x}{4} + t_2\right)^3 + x + \left(-\frac{x}{4} + t_2\right)x - x^3 = 0$$

which simplifies to

$$F_2(x, t_2): \quad 4t_2 + 3t_2^2 + t_2^3 - \frac{t_2 x}{2} - \frac{3t_2^2 x}{4} - \frac{x^2}{16} + \frac{3t_2 x^2}{16} - \frac{65 x^3}{64} = 0.$$

Iteration #2

Step 1: Ignoring all x^n with $n > 2$ in $F_2(x, t_2)$ yields

x^2 level: $4c_2 - \dfrac{1}{16} = 0$

Step 2: $c_2 = \dfrac{1}{64}$.

Step 3: Stop.

The above computations lead to the conclusion that

$$y = 1 - \frac{x}{4} + \frac{x^2}{64} \cdots$$

is an approximate solution to the given equation. More terms can be obtained by further iterations of this procedure.

As is the case for equations with a single unknown, solutions of algebraic equations need not be unique.

Example 4.2.3 Suppose

$$F(x, y) = 3x^2 + 2xy + y^2 + x - y - 2 = 0. \tag{6}$$

This gives,

$F_0(x, t_0): \quad 3x^2 + 2xt_0 + t_0^2 + x - t_0 - 2 = 0.$

Iteration #0

Step 1: Ignoring all the non constant powers of x in $F_0(x, t_0)$, we get

x^0 level: $c_0^2 - c_0 - 2 = 0$

which has solutions $c_0 = -1, 2$. (Accordingly, there will be two solutions for y. We continue here by deriving the solution that corresponds to $y = 1$ and will pursue the other solution below.)

Step 2: $c_0 = -1$ (see comment above)

Step 3: In $F_0(x, t_0)$ replace t_0 with $-1 + t_1$ to obtain

$$3x^2 + 2x(-1 + t_1) + (-1 + t_1)^2 + x - (-1 + t_1) - 2 = 0$$

which simplifies to

$F_1(x, t_1)$: $-3t_1 + t_1^2 - x + 2t_1 x + 3x^2 = 0.$

Iteration #1

Step 1: Ignoring all x^n with $n > 1$ in $F_1(x, t_1)$, we get

x^1 level: $-3c_1 - 1 = 0$

Step 2: $c_1 = -\dfrac{1}{3}$

Step 3: In $F_1(x, t_1)$ replace t_1 with $-x/3 + t_2$ to obtain

$$-3\left(-\frac{x}{3} + t_2\right) + \left(-\frac{x}{3} + t_2\right)^2 - x + 2x\left(-\frac{x}{3} + t_2\right) + 3x^2 = 0$$

which simplifies to

$F_2(x, t_2)$: $-3t_2 + t_2^2 + \dfrac{4t_2 x}{3} + \dfrac{22x^2}{9} = 0$

Iteration #2

Step 1: Ignoring all x^n with $n > 2$ in $F_2(x, t_2)$ we get,

x^2 level: $-3c_2 + \dfrac{22}{9} = 0$

Step 2: $c_2 = \dfrac{22}{27}$

Step 3: Stop.

The above procedure yields

$$y = -1 + \frac{x}{3} + \frac{22x^2}{27} \cdots$$

as a partial solution to the given equation.

On the other hand, if we begin with $c_0 = 2$, the recursive procedure yields:

Iteration #0

Step 1: Same as above

Step 2: $c_0 = 2$

Step 3: In $F_0(x, t_0)$ set $t_0 = 2 + t_1$ to obtain

$$3x^2 + 2x(2 + t_1) + (2 + t_1)^2 + x - (2 + t_1) - 2 = 0$$

which simplifies to

$$F_1(x, t_1): \qquad\qquad 3t_1 + t_1^2 + 5x + 2t_1 x + 3x^2 = 0$$

Iteration #1

Step 1: Ignoring all x^n with $n > 1$ in $F_1(x, t_1)$ we get

x^1 level: $\qquad\qquad 3c_1 + 5 = 0$

Step 2: $c_1 = -\dfrac{5}{3}$

Step 3: In $F_1(x, t_1)$ set $t_1 = -(5x/3) + t_2$ to obtain

$$3\left(-\frac{5x}{3} + t_2\right) + \left(-\frac{5x}{3} + t_2\right)^2 + 5x + 2\left(-\frac{5x}{3} + t_2\right)x + 3x^2 = 0$$

which simplifies to

$$F_2(x, t_2): \qquad\qquad 3t_2 + t_2^2 - \frac{4t_2 x}{3} + \frac{22x^2}{9} = 0.$$

Iteration #2

Step 1: Ignoring all x^n with $n > 2$ in $F_2(x, t_2)$ we get

x^2 level: $\qquad\qquad 3c_2 + \dfrac{22}{9} = 0$

Step 2: $c_2 = -\dfrac{22}{27}$

Step 3: Stop.

Starting with $c_0 = 2$ we obtained

$$y = 2 - \frac{5x}{3} - \frac{22x^2}{27} \cdots$$

as a partial solution to the given equation.

Because Equation (6) is of the second degree in y there is another way to derive the infinite series expansion of y. This equation can be solved for y by applying the quadratic formula and then going on to evaluate the radical in the solution by means of the Fractional Binomial Theorem (Exercise 3).

As was the case for the differential equations of the previous section, the method described here needs further amplification (see Exercise 6). The details again fall outside the scope of this text.

Exercises 4.2

1. Find the first 3 terms of the infinite series expansion of y in terms of x in the equations below in two different ways: using the method described in this section, and using the Fractional Binomial Theorem.

a) $y - xy = 1$ b) $y^2 - x - 1 = 0$
c) $y^2 - (1 + x)^3 = 0$ d) $y^3 - (1 + x)^2 = 0$
e) $y - 2xy + x^2 y = 1$ f) $y - x^2 y = 1$

2. Without the use of a calculating device, find the first 3 terms of the expansion of y in terms of x where the two variables are related by:

a) $y - xy = 0$ b) $y^3 + y + x = 0$
c) $y^3 + xy = 1$ d) $y^2 + y + x = 0$
e) $x - 2y + y^2 = 0$ f) $y^2 - y = x$
g) $y^3 + 2y - 3 + xy - x^3 = 0$ h) $y^3 + 3y - 4 - x^2 y + 2x = 0$
i) $x^2 y^2 - 2y^3 + x - y + 18 = 0$ j) $x^2 + 2xy - 2y^2 + x - y + 1 = 0$

3. Solve Equation (6) for y by means of the quadratic equation and then use the fractional version of the Binomial Theorem to find the first three terms of the power series expansion of both solutions.

4. Use a symbol manipulator to derive the first 5 terms of the expansion of y in terms of x for the equations listed in Exercise 2.

5. Use a symbol manipulator to derive the first 5 terms of the expansion of y in terms of x where they are related by:

a) $x^2 - 3xy^3 + y^4 - 16 = 0$ b) $x^2 - 3xy^3 + y^5 - 1 = 0$
c) $x^2 - 3xy^3 + y^6 - 1 = 0$ d) $x^2 - 3xy^3 + y^7 - 1 = 0$.

6. Show that this section's method fails to solve the equation $xy = 1$. Explain how this is remedied by the substitution $u = 1 + x$.

7. Find the first 5 terms of the infinite series expansion of the y in terms of x in the equations below.

a) $y^3 + y = 2 \cos x$ (Hint: Use Exercise 4.1.4)
b) $y^3 - y = \sin x$ (Hint: Use Exercise 4.1.3)
c) $y = \tan x$ (Hint: Use Equation (12) of Chapter 3)

d) $\sin y = x$ (Hint: Use Exercise 4.1.3)
e) $e^y = 1 + x$ (Hint: Use Exercise 4.1.5)

8. Let $P(x)$ be a polynomial in x such that $P(0)$ is positive. Prove that this section's method will produce a solution to the equation $y^2 = P(x)$.

Chapter Appendix: Mathematica implementations of Newton's algorithm

Example 4.2.1 $y^3 - x - 1 = 0$

```
In[31]: =  Original = y^3 - 1 - x
Out[31] = -1 - x + y^3
In[32]: =  F0 = Original //.y -> t0
Out[32] = -1 + t0^3 - x
In[33]: =  Solve[(F0//.x -> 0) == 0,t0]
Out[33] = {{t0 → 1}, {t0 → -(-1)^(1/3)}, {t0 → (-1)^(2/3)}}
In[34]: =  F1 = Expand[F0//.t0 -> 1+t1]
Out[34] = 3t1 + 3t1^2 + t1^3 - x
In[35]: =  Solve[3t1-x == 0, t1]
Out[35] = {{t1 → x/3}}
In[36]: =  F2 = Expand[F1//.t1 -> x/3+t2]
```

$Out[36] = 3t2 + 3t2^2 + t2^3 + 2t2x + t2^2 x + \frac{x^2}{3} + \frac{t2x^2}{3} + \frac{x^3}{27}$

```
In[37]: =  Solve[3t2+x^2/3 == 0,t2]
```

$Out[37] = \left\{\left\{t2 \to -\frac{x^2}{9}\right\}\right\}$

```
In[38]: =  F3 = Expand[F2//.t2 -> -x^2/9+t3]
```

$Out[38] = 3t3+3t3^2+t3^3+2t3x+t3^2 x-\frac{t3x^2}{3}-\frac{t3^2 x^2}{3}-\frac{5x^3}{27}-\frac{2t3x^3}{9}+\frac{t3x^4}{27}+\frac{x^5}{81}-\frac{x^6}{729}$

```
In[39]: =  Solve[3t3-5x^3/27 == 0,t3]
```

$Out[39] = \left\{\left\{t3 \to \frac{5x^3}{81}\right\}\right\}$

```
In[40]: =  F4 = Expand[F3//.t3 -> 5x^3/81+t4]
```

$Out[40] = 3t4 + 3t4^2 + t4^3 + 2t4x + t4^2 x - \frac{t4x^2}{3} - \frac{t4^2 x^2}{3} + \frac{4t4x^3}{27} + \frac{5t4^2 x^3}{27} + \frac{10x^4}{81} +$
$\frac{13t4x^4}{81} - \frac{2x^5}{243} - \frac{10t4x^5}{243} - \frac{8x^6}{2187} + \frac{25t4x^6}{2187} + \frac{40x^7}{6561} - \frac{25x^8}{19683} + \frac{125x^9}{531441}$

```
In[41]: =  Solve[3t4+10x^4/8 == 0,t4]
```

$Out[41] = \left\{\left\{t4 \to -\frac{10x^4}{243}\right\}\right\}$

Hence

$$y = 1 + \frac{x}{3} - \frac{x^2}{9} + \frac{5x^3}{81} - \frac{10x^4}{243} \cdots$$

Example 4.2.2 $y^3 + y - 2 + xy - x^3 = 0$

```
In[15]: =  Original = y^3+y-2+x*y-x^3
Out[15] = -2 - x^3 + y + xy + y^3
In[16]: =  F0 = Original //.y -> t0
Out[16] = -2 + t0 + t0^3 + t0x - x^3
In[22]: =  Solve[(F0//.x -> 0) == 0,t0]
```

$Out[22] = \left\{\{t0 \to 1\}, \left\{t0 \to \frac{1}{2}(-1 - I\sqrt{7})\right\}, \left\{t0 \to \frac{1}{2}(-1 + I\sqrt{7})\right\}\right\}$

```
In[23]: =  F1 = Expand[F0//.t0 -> 1+t1]
Out[23] = 4t1 + 3t1^2 + t1^3 + x + t1x - x^3
In[24]: =  Solve[4t1+x == 0, t1]
```

$Out[24] = \left\{\left\{t1 \to -\frac{x}{4}\right\}\right\}$

```
In[25]: =  F2 = Expand[F1//.t1 -> (-x/4)+t2]
```

$Out[25] = 4t2 + 3t2^2 + t2^3 - \frac{t2x}{2} - \frac{3t2^2x}{4} - \frac{x^2}{16} + \frac{3t2x^2}{16} - \frac{65x^3}{64}$

$In[26]: = $ **Solve[4t2-x^2/16 == 0,t2]**

$Out[26] = \left\{\left\{t2 \to \frac{x^2}{64}\right\}\right\}$

$In[27]: = $ **F3 = Expand[F2//.t2 -> (x^2)/64+t3]**

$Out[27] = 4t3 + 3t3^2 + t3^3 - \frac{t3x}{2} - \frac{3t3^2x}{4} + \frac{9t3x^2}{32} + \frac{3t3^2x^2}{64} - \frac{131x^3}{128} - \frac{3t3x^3}{128} + \frac{15x^4}{4096} + \frac{t3x^4}{4096} - \frac{3x^5}{16384} + \frac{x^6}{262144}$

$In[28]: = $ **Solve[4t3+131x^3/128 == 0,t3]**

$Out[28] = \left\{\left\{t3 \to -\frac{131x^3}{512}\right\}\right\}$

$In[29]: = $ **F4 = Expand[F3//.t3 -> -131x^3/128+t4]**

$Out[29] = 4t4 + 3t4^2 + t4^3 - \frac{t4x}{2} - \frac{3t4^2x}{4} - \frac{9t4x^2}{32} + \frac{3t4^2x^2}{64} - \frac{655x^3}{128} - \frac{789t4x^3}{128} - \frac{393t4^2x^3}{128} + \frac{2111x^4}{4096} + \frac{629t4x^4}{4096} - \frac{4719x^5}{16384} - \frac{393t4x^5}{4096} + \frac{830017x^6}{262144} + \frac{51483t4x^6}{16384} - \frac{412257x^7}{524288} + \frac{51483x^8}{1048576} - \frac{2248091x^9}{2097152}$

$In[30]: = $ **Solve[4t4+2111x^4/4096 == 0,t4]**

$Out[30] = \left\{\left\{t4 \to -\frac{2111x^6}{16384}\right\}\right\}$

Hence

$$y = 1 - \frac{x}{4} + \frac{x^2}{64} - \frac{131x^3}{512} - \frac{2111x^4}{16384} \cdots$$

Example 4.2.3 $3x^2 + 2xy + y^2 + x - y - 2 = 0$

$In[42]: = $ **Original = 3x^2+2x*y+y^2+x-y-2**

$Out[42] = -2 + x + 3x^2 - y + 2xy + y^2$

$In[43]: = $ **F0 = Original //.y -> t0**

$Out[43] = -2 - t0 + t0^2 + x + 2t0x + 3x^2$

$In[44]: = $ **Solve[(F0//.x -> 0) == 0,t0]**

$Out[44] = \{\{t0 \to 1\}, \{t0 \to 2\}\}$

(* First branch *)

$In[45]: = $ **F1 = Expand[F0//.t0 ->-1+t1]**

$Out[45] = -3t1 + t1^2 - x + 2t1x + 3x^2$

$In[46]: = $ **Solve[-3t1-x == 0, t1]**

$Out[46] = \left\{\left\{t1 \to -\frac{x}{3}\right\}\right\}$

$In[47]: = $ **F2 = Expand[F1//.t1 -> -x/3+t2]**

$Out[47] = -3t2 + t2^2 + \frac{4t2x}{3} + \frac{22x^2}{9}$

$In[49]: = $ **Solve[-3t2+22x^2/9 == 0,t2]**

$Out[49] = \left\{\left\{t2 \to \frac{22x^2}{27}\right\}\right\}$

$In[51]: = $ **F3 = Expand[F2//.t2 -> 22x^2/27+t3]**

$Out[51] = -3t3 + t3^2 + \frac{4t3x}{3} + \frac{44t3x^2}{27} + \frac{88x^3}{81} + \frac{484x^4}{729}$

$In[57]: = $ **Solve[-3t3+88x^3/81 == 0,t3]**

$Out[57] = \left\{\left\{t3 \to \frac{88x^3}{243}\right\}\right\}$

Hence

$$y = -1 - \frac{x}{3} + \frac{22x^2}{27} - \frac{88x^3}{243} \cdots$$

(* Second branch *)

$In[52]: = $ **F1 = Expand[F0//.t0 ->2+t1]**

$Out[52] = 3t1 + t1^2 + 5x + 2t1x + 3x^2$

$In[53]: = $ **Solve[3t1+5x == 0, t1]**

$Out[53] = \left\{\left\{t1 \to -\frac{5x}{3}\right\}\right\}$

$In[54]: = $ **F2 = Expand[F1//.t1 -> -5x/3+t2]**

$Out[54] = 3t2 + t2^2 - \frac{4t2x}{3} + \frac{22x^2}{9}$

In[55]: = **Solve[3t2+22x^2/9 == 0,t2]**

Out[55] = $\left\{\left\{t2 \to -\frac{22x^2}{27}\right\}\right\}$

In[56]: = **F3 = Expand[F2//.t2 -> 22x^2/27+t3]**

Out[56] = $3t3 + t3^2 - \frac{4t3x}{3} - \frac{44t3x^2}{27} + \frac{88x^3}{81} + \frac{484x^4}{729}$

In[58]: = **Solve[3t3+88x^3/81 == 0,t3]**

Out[58] = $\left\{\left\{t3 \to -\frac{88x^3}{243}\right\}\right\}$

Hence

$$y = 2 - \frac{5x}{3} - \frac{22x^2}{27} - \frac{88x^3}{243} \cdots$$

Chapter Summary

Mathematics progresses by posing problems and finding their solutions. Newton's power series above and beyond providing methods for evaluating complicated functions also make possible the effective solution of interesting differential and algebraic equations.

5

Euler

The most dominant figures on the mathematical landscape of the 18th century were Leonhard Euler(1707–1783) and Joseph Louis Lagrange (1736–1813). One of their many contributions to calculus is their recognition of the importance of infinite series of trigonometric functions to both pure and applied mathematics. The discussion below, extracted from an article of Euler's, has several interesting aspects: complex numbers are used to obtain new information about real numbers, and manifestly false equations lead to correct and interesting results.

5.1 TRIGONOMETRIC SERIES

In his article *Subsiduum Calculi Sinnum*, Euler put the infinite geometric progression

$$\frac{1}{1-x} = 1 + x + x^2 + x^3 + \cdots$$

to a very surprising use. He replaced the variable x by the complex number

$$z = a(\cos\phi + i\sin\phi)$$

where $i = \sqrt{-1}$ to obtain the series

$$\frac{1}{1 - a(\cos\phi + i\sin\phi)} = 1 + [a(\cos\phi + i\sin\phi)]$$
$$+ [a(\cos\phi + i\sin\phi)]^2 + [a(\cos\phi + i\sin\phi)]^3 \cdots . \quad (1)$$

Euler discussed the case $a = -1$ in some detail and we shall restrict attention to it immediately. Applying standard procedures for rationalizing denominators, the left-side of Equation (1) is transformed as follows:

$$\frac{1}{1 + (\cos\phi + i\sin\phi)} = \frac{1}{1 + \cos\phi + i\sin\phi} \cdot \frac{1 + \cos\phi - i\sin\phi}{1 + \cos\phi - i\sin\phi}$$
$$= \frac{1 + \cos\phi - i\sin\phi}{(1 + \cos\phi)^2 + \sin^2\phi} = \frac{1 + \cos\phi - i\sin\phi}{2(1 + \cos\phi)}$$
$$= \frac{1}{2} - \frac{i\sin\phi}{2(1 + \cos\phi)} \tag{2}$$

On the right-side of Equation (1), De Moivre's Theorem

$$(\cos\phi + i\sin\phi)^k = \cos k\phi + i\sin k\phi$$

can be used to obtain

$$1 - (\cos\phi + i\sin\phi) + (\cos\phi + i\sin\phi)^2 - (\cos\phi + i\sin\phi)^3 \cdots$$
$$= 1 - (\cos\phi + i\sin\phi) + (\cos 2\phi + i\sin 2\phi) - (\cos 3\phi + i\sin 3\phi) \cdots$$
$$= (1 - \cos\phi + \cos 2\phi - \cos 3\phi + \cdots) - i(\sin\phi - \sin 2\phi + \sin 3\phi \cdots). \tag{3}$$

Equating the real parts of Equations (2) and (3) Euler obtained the equation

$$\frac{1}{2} = 1 - \cos\phi + \cos 2\phi - \cos 3\phi + \cos 4\phi - \cos 5\phi \cdots$$

or

$$\cos\phi - \cos 2\phi + \cos 3\phi - \cos 4\phi + \cos 5\phi \cdots = \frac{1}{2}. \tag{4}$$

When this equation is integrated, with integration constant C_1, we get

$$\sin\phi - \frac{1}{2}\sin 2\phi + \frac{1}{3}\sin 3\phi - \frac{1}{4}\sin 4\phi + \frac{1}{5}\sin 5\phi \cdots = \frac{\phi}{2} + C_1. \tag{5}$$

Since $\sin 0 = 0$, the substitution $\phi = 0$ yields $C_1 = 0$, so that another integration results in

$$\cos\phi - \frac{1}{2^2}\cos 2\phi + \frac{1}{3^2}\cos 3\phi - \frac{1}{4^2}\cos 4\phi + \frac{1}{5^2}\cos 5\phi \cdots = C_2 - \frac{\phi^2}{4}. \tag{6}$$

Since $\cos 0 = 1$, the substitution $\phi = 0$ yields

$$C_2 = 1 - \frac{1}{2^2} + \frac{1}{3^2} - \frac{1}{4^2} + \frac{1}{5^2} \cdots . \tag{7}$$

The right-side of Equation (7) had been studied by Euler (and others) before, and he already knew that it actually equals $\pi^2/12$ and in the article under discussion Euler merely referred to this fact in order to evaluate the integration constant C_2 in Equation (6). Here, instead, this interesting fact will be proved by means of an argument that, while not rigorous by today's standards, would certainly have been accepted by Euler and his contemporaries.

Proposition 5.1.1 $\dfrac{\pi^2}{12} = 1 - \dfrac{1}{2^2} + \dfrac{1}{3^2} - \dfrac{1}{4^2} + \dfrac{1}{5^2} \cdots$

Proof. Recall that according to Equation (7)

$$C_2 = 1 - \frac{1}{2^2} + \frac{1}{3^2} - \frac{1}{4^2} + \frac{1}{5^2} \cdots$$

On the other hand, bearing in mind that $\cos(2\pi + \alpha) = \cos \alpha$, the substitution of $\phi = \pi/2$ into Equation (6) yields

$$C_2 - \frac{(\pi/2)^2}{4} = \cos \frac{\pi}{2} - \frac{1}{2^2} \cos \pi + \frac{1}{3^2} \cos \frac{3\pi}{2} - \frac{1}{4^2} \cos 2\pi + \frac{1}{5^2} \cos \frac{\pi}{2} \cdots$$

or

$$C_2 - \frac{\pi^2}{16} = 0 - \frac{1}{2^2}(-1) + 0 - \frac{1}{4^2} + 0 - \frac{1}{6^2}(-1) + 0 - \frac{1}{8^2} \cdots$$
$$= \frac{1}{4} \left(1 - \frac{1}{2^2} + \frac{1}{3^2} - \frac{1}{4^2} \cdots \right) = \frac{1}{4} C_2.$$

Solving for C_2 now yields

$$C_2 = \frac{\dfrac{\pi^2}{16}}{\dfrac{3}{4}} = \frac{\pi^2}{12} \qquad\qquad \square$$

The above proposition begs the question of evaluating

$$1 + \frac{1}{2^2} + \frac{1}{3^2} + \frac{1}{4^2} + \cdots .$$

This problem had also been considered by several of Euler contemporaries and was eventually solved by Euler in a manner different from that presented here. This particular series appears in several surprising contexts. For example, it yields the probability that two integers chosen at random are relatively prime (see Proposition 9.4.2 below).

Proposition 5.1.2 $\dfrac{\pi^2}{6} = 1 + \dfrac{1}{2^2} + \dfrac{1}{3^2} + \dfrac{1}{4^2} + \cdots .$

Proof. By Proposition 5.1.1,

$$\frac{\pi^2}{12} = 1 - \frac{1}{2^2} + \frac{1}{3^2} - \frac{1}{4^2} + \frac{1}{5^2} - \frac{1}{6^2} + \frac{1}{7^2} - \frac{1}{8^2} \cdots$$

$$= \left(1 + \frac{1}{2^2} + \frac{1}{3^2} + \frac{1}{4^2} + \frac{1}{5^2} + \frac{1}{6^2} + \frac{1}{7^2} + \frac{1}{8^2} \cdots \right)$$
$$- 2\left(\frac{1}{2^2} + \frac{1}{4^2} + \frac{1}{6^2} + \frac{1}{8^2} \cdots \right)$$
$$= \left(1 + \frac{1}{2^2} + \frac{1}{3^2} + \frac{1}{4^2} + \frac{1}{5^2} + \frac{1}{6^2} + \frac{1}{7^2} + \frac{1}{8^2} \cdots \right)$$
$$- \frac{2}{2^2}\left(1 + \frac{1}{2^2} + \frac{1}{3^2} + \frac{1}{4^2} \cdots \right)$$
$$= \frac{1}{2}\left(1 + \frac{1}{2^2} + \frac{1}{3^2} + \frac{1}{4^2} + \frac{1}{5^2} + \frac{1}{6^2} + \frac{1}{7^2} + \frac{1}{8^2} \cdots \right).$$

It follows that

$$\frac{\pi^2}{6} = 1 + \frac{1}{2^2} + \frac{1}{3^2} + \frac{1}{4^2} + \cdots . \qquad \square$$

It may be felt that the content of Propositions 5.1.1 and 5.1.2, while perhaps interesting, hardly deserve the status of a *proposition*, since they seem to be lacking in generality. It is, however, common practice amongst mathematicians to label interesting observations as propositions, or theorems. In this particular case, the significance of these formulas is enhanced by the fact that the complex function which is defined indirectly by

$$\zeta(s) = 1 + \frac{1}{2^s} + \frac{1}{3^s} + \frac{1}{4^s} + \cdots$$

and which for $s = 2$ yields the sum evaluated in Proposition 5.1.2, contains much information regarding the distribution of prime numbers. So much so that the nature of the solutions of the equation

$$\zeta(s) = 0$$

is considered by many to be the most important unresolved problem of pure mathematics (see Section 9.4 for some more details). Of course, such solutions have to be imaginary numbers, but that does not detract from their importance.

Euler's aforementioned algebraic manipulations are interesting because they also bring to the foreground a new type of infinite series. To explain this it is best to reexamine the power series of the previous chapters.

The assumption that underlies Newton's work is that every function can be expressed as a power series of the type $c_0 + c_1x + c_2x^2 + c_3x^3 + \cdots$, and that these series by and large behave like the polynomials they resemble. This assumption was justified solely on the grounds of its utility. It made possible the effective solution of otherwise unsolvable problems, and since these solutions did not seem to lead to any outrageous contradictions, the assumption might as well be taken for granted.

Just like polynomials, trigonometric functions also have properties that make them particularly useful under certain circumstances. The most prominent of these are the periodicity $f(x+2\pi) = f(x)$ enjoyed by all trigonometric functions and the periodicities of the derivatives of the $\sin x$ and $\cos x$ functions, namely that these two functions satisfy the differential equation

$$y'' = -y \qquad (8)$$

and, consequently,

$$y^{(\text{iv})} = y.$$

It so happens that the periodicity exhibited in Equation (8) turns the trigonometric functions into the building blocks of the solutions to the two partial differential equations

$$k^2\frac{\partial^2 f}{\partial x^2} = \frac{\partial f}{\partial t} \quad \text{and} \quad k^2\frac{\partial^2 f}{\partial x^2} = \frac{\partial^2 f}{\partial t^2} \qquad (9)$$

where $f = f(x,t)$ and k is a positive constant. Equations (9) are known as the *heat equation* and the *wave equation* respectively. They are of fundamental importance in physics since, as their names imply, they govern the propagation of heat, sound, and light. A detailed exposition of the genesis of these equations and the role they played in the evolution of the concept of a mathematical function can be found in the references. At this point, suffice it to say that it is easily seen that for each positive integer n the function

$$f(x,t) = \sin nx \cos nkt$$

constitutes a solution of the wave equation of Equation (9), as is every combination of the form

$$a_1 \sin x \cos kt + a_2 \sin 2x \cos 2kt + \cdots + a_n \sin nx \cos nkt. \qquad (10)$$

Given the example set by Newton's infinite series, it is not surprising that the mathematicians working on the wave equations assumed that the infinite extensions of Equation (10), namely that infinite trigonometric series of the type

$$a_1 \sin x \cos kt + a_2 \sin 2x \cos 2kt + a_3 \sin 3x \cos 3kt + \cdots$$

also constitute solutions of the wave equation. The question of whether these infinite trigonometric series actually described all the solutions of the wave equation then devolved to the question of

What functions $f(x)$ can be expressed in one of the forms

$$a_0 + a_1 \sin x + a_2 \sin 2x + a_3 \sin 3x + \cdots \qquad (11)$$

or

$$a_0 + a_1 \cos x + a_2 \cos 2x + a_3 \cos 3x + \cdots? \qquad (12)$$

The computations preceding Proposition 5.1.1 are tantamount to showing that the two functions $f(x) = x$ and $f(x) = x^2$ can indeed be expressed in these forms. However, some qualification is required here. Since

$$\sin(x + 2\pi) = \sin x \quad \text{and, in general,} \quad \sin n(x + 2\pi) = \sin nx$$

it follows that any function $f(x)$ that is expressible in the forms of Equations (11) and (12) must also possess the periodicity $f(x + 2\pi) = f(x)$. However, x and x^2, like most functions, do not possess this periodicity. This difficulty is circumvented by restricting attention to the domain $(-\pi, \pi)$ which is the largest open interval to which the 2π periodicity does not apply.

Proposition 5.1.3 *For* $x \in (-\pi, \pi)$

$$x = 2 \left(\sin x - \frac{\sin 2x}{2} + \frac{\sin 3x}{3} - \frac{\sin 4x}{4} + \cdots \right).$$

Proof. This follows immediately from Equation (5) above and the subsequent observation that $C_1 = 0$. □

Proposition 5.1.4 *For* $x \in (-\pi, \pi)$

$$x^2 = 4 \left(\frac{\pi^2}{12} - \cos x + \frac{\cos 2x}{2^2} - \frac{\cos 3x}{3^2} + \frac{\cos 4x}{4^2} - \cdots \right).$$

Proof. This follows from Equations (6) and (7) and Proposition 5.1.1. □

All the proofs of this section should be taken with a healthy dose of skepticism. Despite his unquestioned status as one of the greatest mathematicians of all times, Euler is notorious for producing faulty proofs and making demonstrably false assertions. Euler's proofs of Propositions 5.1.1–4 fall in the first of these two categories. In other words, while these propositions' assertions are valid, Euler's proofs are incomplete. Valid proofs can be found in Section 14.1 and elsewhere.

Euler's work on the trigonometric series was mostly ignored by the mathematicians of the 18th century. Interest was revived in 1807 when Joseph Fourier (1768–1830) presented a paper before the French Academy of Sciences in which he pointed out the relevance of such series to the propagation of heat. A revised version of the paper won a prize from the Academy. However, because Fourier's arguments were just as flawed as Euler's, the Academy refused to publish his paper in its *Mémoires*, thus underscoring the mathematical establishment's ambivalence about the need for rigor. Only after Fourier's appointment as secretary of the Academy was he able to publish his award winning paper in the Academy's *Mémoires*. The subject matter eventually grew into a major branch of mathematics now known as *Fourier Analysis*.

Exercises 5.1

1. Use Equation (1) to derive the formulas:

a) $\dfrac{1 - a\cos\phi}{1 - 2a\cos\phi + a^2} = 1 + a\cos\phi + a^2\cos 2\phi + a^3\cos 3\phi \cdots$

b) $\dfrac{\sin\phi}{1 - 2a\cos\phi + a^2} = \sin\phi + a\sin 2\phi + a^2\sin 3\phi + a^3\sin 4\phi \cdots.$

2. Using a calculator compare the graphs of

$$f_1(x) = \frac{1 - a\cos x}{1 - 2a\cos x + a^2}$$

and

$$f_2(x) = 1 + a\cos x + a^2\cos 2x + \cdots + a^8\cos 8x$$

for $-\pi < x < \pi$ and the following values of a:

a)	$a = .5$	b)	$a = -.5$	c)	$a = .7$
d)	$a = -.7$	e)	$a = .9$	f)	$a = -.9$
g)	$a = 1$	h)	$a = -1.$		

3. Using a calculator compare the graphs of

$$f_1(x) = \frac{\sin x}{1 - 2a\cos x + a^2}$$

and

$$f_2(x) = \sin x + a\sin 2x + \cdots + a^7\sin 8x$$

for $-\pi < x < \pi$ and the following values of a:

a)	$a = .5$	b)	$a = -.5$	c)	$a = .7$
d)	$a = -.7$	e)	$a = .9$	f)	$a = -.9$
g)	$a = 1$	h)	$a = -1.$		

4. a) Show that

$$\frac{x^3}{12} - \frac{\pi^2 x}{12} = -\sin x + \frac{\sin 2x}{2^3} - \frac{\sin 3x}{3^3} + \frac{\sin 4x}{4^3} - \cdots \qquad \text{for } -\pi < x < \pi$$

b) Show that

$$\frac{x^4}{48} - \frac{\pi^2 x^2}{24} + \frac{7\pi^4}{720} = \cos x - \frac{\cos 2x}{2^4} + \frac{\cos 3x}{3^4} - \frac{\cos 4x}{4^4} + \cdots \qquad \text{for } -\pi < x < \pi$$

c) Show that

$$\frac{\pi^4}{90} = 1 + \frac{1}{2^4} + \frac{1}{3^4} + \frac{1}{4^4} + \cdots$$

d) Show that

$$\frac{\pi^6}{945} = 1 + \frac{1}{2^6} + \frac{1}{3^6} + \frac{1}{4^6} + \cdots$$

5. Using a graphing calculator compare the graphs of $f_1(x) = x$ and

$$f_2(x) = 2\left(\sin x - \frac{\sin 2x}{2} + \frac{\sin 3x}{3} - \cdots + (-1)^{n-1}\frac{\sin nx}{n}\right)$$

for $-\pi < x < \pi$ and for the following values of n:
   a) $n = 10$ b) $n = 20$ c) $n = 30$ d) $n = 40$.

6. Using a graphing calculator compare the graphs of $f_1(x) = x^2$ and

$$f_2(x) = 4\left(\frac{\pi^2}{12} - \cos x + \frac{\cos 2x}{4} - \frac{\cos 3x}{9} + \cdots + (-1)^n\frac{\cos nx}{n^2}\right)$$

for $-\pi < x < \pi$ and for the following values of n:
   a) $n = 10$ b) $n = 20$ c) $n = 30$ d) $n = 40$.

7. Substitute the values below for x in Equation (4). Is the convergence valid?
   a) 0   b) $\pi/4$ c) $\pi/3$ d) $\pi/2$ e) π.

8. Use a calculator to evaluate $\cos x - \cos 2x + \cos 3x - \cos 4x + \cdots + (-1)^{n-1}\cos nx$ for $n = 10, 30, 50, 70$ each where x equals
   a) $10°$ b) $20°$ c) $30°$ d) $40°$ e) $50°$
   What does your answer say about Equation (4)?

9. Prove that $\dfrac{\pi}{3\sqrt{3}} = 1 - \dfrac{1}{2} + \dfrac{1}{4} - \dfrac{1}{5} + \dfrac{1}{7} - \dfrac{1}{8} + \dfrac{1}{10} - \dfrac{1}{11} + \cdots$

10. Prove that $\dfrac{2\pi}{3\sqrt{3}} = 1 + \dfrac{1}{2} - \dfrac{1}{4} - \dfrac{1}{5} + \dfrac{1}{7} + \dfrac{1}{8} - \dfrac{1}{10} - \dfrac{1}{11} + \cdots$

11. Prove that $\dfrac{\pi}{2\sqrt{2}} = 1 + \dfrac{1}{3} - \dfrac{1}{5} - \dfrac{1}{7} + \dfrac{1}{9} + \dfrac{1}{11} - \dfrac{1}{13} - \dfrac{1}{15} \cdots$

12. Prove that $\dfrac{4\pi^2}{27} = 1 + \dfrac{1}{2^2} + \dfrac{1}{4^2} + \dfrac{1}{5^2} + \dfrac{1}{7^2} + \dfrac{1}{8^2} + \dfrac{1}{10^2} + \dfrac{1}{11^2} + \cdots$

13. Prove that $\dfrac{2\pi^2}{27} = 1 - \dfrac{1}{2^2} - \dfrac{1}{4^2} + \dfrac{1}{5^2} + \dfrac{1}{7^2} - \dfrac{1}{8^2} - \dfrac{1}{10^2} + \dfrac{1}{11^2} + \cdots$

14. Prove that $\dfrac{\pi^2}{8} = 1 + \dfrac{1}{3^2} + \dfrac{1}{5^2} + \cdots$

15. Prove that $\dfrac{\pi^2}{8\sqrt{2}} = 1 - \dfrac{1}{3^2} - \dfrac{1}{5^2} + \dfrac{1}{7^2} + \dfrac{1}{9^2} - \dfrac{1}{11^2} - \dfrac{1}{13^2} + \cdots$

16. Prove that $\dfrac{\pi^3}{32} = 1 - \dfrac{1}{3^3} + \dfrac{1}{5^3} - \cdots$

17. Express $1 + \dfrac{1}{3^4} + \dfrac{1}{5^4} + \cdots$ in terms of π.

18. Express $a + \dfrac{b}{2^2} + \dfrac{a}{3^2} + \dfrac{b}{4^2} + \dfrac{a}{5^2} + \dfrac{b}{6^2} + \cdots$ in terms of a, b, and π.

19. Express $1 + \dfrac{1}{2^2} - \dfrac{1}{3^2} + \dfrac{1}{4^2} + \dfrac{1}{5^2} - \dfrac{1}{6^2} + \dfrac{1}{7^2} + \dfrac{1}{8^2} - \dfrac{1}{9^2} \cdots$ in terms of π.

20. Use Proposition 5.1.3 to prove that $\dfrac{\pi}{4} = 1 - \dfrac{1}{3} + \dfrac{1}{5} - \dfrac{1}{7} \pm \cdots$

21. Express the series $1 - \dfrac{1}{7} + \dfrac{1}{9} - \dfrac{1}{15} + \dfrac{1}{17} - \dfrac{1}{23} + \dfrac{1}{25} \mp \cdots$ in terms of π.

22. Describe the main mathematical achievements of
    a) Leonhard Euler
    b) Joseph Fourier
    c) Joseph Louis Lagrange

Chapter Summary

In the mid eighteenth century Euler applied the infinite geometrical series to complex numbers and showed how to express functions of the form x^n as infinite series of trigonometric functions. Such trigonometric series are very convenient when describing solutions of both the heat and wave equations of physics. Their study eventually gave rise to several new mathematical disciplines. However, the strange properties of these series also point out the need for extreme caution in integrating and differentiating infinite series. This is particularly well exemplified by the following three equations derived above by Euler

$$\frac{1}{2} = \cos x - \cos 2x + \cos 3x - \cos 4x \cdots$$

$$\frac{x}{2} = \sin x - \frac{\sin 2x}{2} + \frac{\cos 3x}{3} - \frac{\sin 4x}{4} \cdots$$

$$\frac{x^2}{4} = \frac{\pi^2}{12} - \cos x + \frac{\cos 2x}{2^2} - \frac{\cos 3x}{3^2} + \frac{\cos 4x}{4^2} \cdots$$

The last two are valid for $-\pi < x < \pi$ (see Exercises 5, 6 and Propositions 14.1.2, 14.1.4) whereas the first is false for *all* values of x (see Exercises 7, 8, Example 9.1.4, and Exercise 9.1.3). Yet it is the same procedure of differentiating infinite series that leads from the valid last equation to the valid middle equation and from there to the *false* first equation.

$$6$$

The Real Numbers

The remaining chapters of this book are dedicated to the task of replacing most of the previous unrigorous arguments with acceptable proofs. This calls for a careful reexamination of several fundamental concepts. This chapter contains basic information about the real numbers system, rational and irrational numbers, sets, and functions.

6.1 AN INFORMAL INTRODUCTION

It is both natural and customary to picture numbers on a horizontal straight line, the so called *number line*. If the number r is represented by the point P_r then this representation has two properties:

If $r < s$ then P_r falls to the left of P_s;

For each number r, $\frac{P_r P_0}{P_1 P_0} = r$.

Since it is always possible to find a new rational number between any two rational numbers, it was natural for the mathematicians of antiquity to assume that the points representing the rational numbers actually filled out the number line. The falsity of this assumption seems to have come as a surprise to the early Greek mathematicians and the bearer of these unpleasant tidings was reportedly drowned. This tale may be apocryphal. Certainly by the time Euclid wrote his celebrated book *Elements* (circa 300 B.C.) Greek mathematicians had developed a theory of numbers that fully accounted for all the points on the number line. Nevertheless, the fact that proofs and discussions of the existence of non-rational points on the number line can be found in the writings of both Plato (427–347 B.C.) and Aristotle (384–322 B.C.) attests to the impact that this discovery exerted on the first pure mathematicians.

The Greeks stated their discovery of irrational numbers geometrically. Specifically, they proved that if X is that point on the number line for which $P_0 X$ has the same length as the diagonal of the square with side $P_0 P_1$, then there is no rational number r such that $X = P_r$ (Fig. 6.1). This fact is now rephrased and proved in more modern terms.

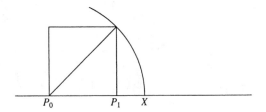

Fig. 6.1   The square root of 2.

Proposition 6.1.1 *There is no rational number r such that*

$$r^2 = 2.$$

Proof. Suppose to the contrary that there is such a rational number $r = m/n$ and that this fraction is expressed in lowest terms, that is, m and n are integers with no common prime factor. It then follows that

$$m^2 = 2n^2$$

and, in particular, m^2 is an even integer. Since the square of every even integer is even and the square of every odd integer is odd, m itself must be even. In other words, there is an integer a such that $m = 2a$. Substitution into the above equation yields

$$(2a)^2 = 2n^2 \quad \text{or} \quad 2a^2 = n^2$$

from which we again conclude that n is even. This, however, contradicts our assumption that the fraction m/n is in lowest terms. Consequently, there is no rational number m/n whose square is 2. □

In all honesty it should be pointed out that this proof is incomplete because it makes use of properties of integers that require proofs. Specifically, it was assumed that if the integer power m^k is divisible by some prime number p, then so is m divisible by p. These properties, however, are both valid and intuitively plausible and so we shall not take the time to fill in the details here. An alternate and complete proof of the non-rationality of $\sqrt{2}$ can be found in Section 6.4.

The Greeks, as well as every other culture that considered such matters, took it for granted that every line segment has a numerical length; in other words, that every point on the number line represents some number. Those numbers that corresponded to points not covered by the rational numbers, such as the point X in Figure 6.1, became known as the *irrational* numbers. Their description at first gave rise to some difficulties, but these difficulties were eventually resolved by the decimal number system. For our purposes here a decimal number has the form $\pm N.d_1 d_2 d_3 \ldots$ where N is a non negative integer and $d_1, d_2, d_3, \ldots$ are any of the digits $0, 1, 2, \ldots, 9$. By the time calculus was invented the following facts were taken for granted by all mathematicians:

For every point P on the number line there is a decimal number r such that $P = P_r$;

For every decimal number r there is a point P on the number line such that $P = P_r$.

These decimal numbers eventually became known as the real numbers in order to distinguish them from the so called imaginary numbers that emerged in the sixteenth century.

While the identification of the real numbers with the decimal expansions provides us with a practical method for calculating with real numbers, it suffers from two defects. It seems improper to describe a natural phenomenon such as the straight line in terms of a human construct that is based on the coincidence that earthlings have ten fingers. Moreover, the technical details of the decimal system and its occasional ambiguities result in considerable complications when it comes to firming up the theoretical foundations of calculus. For this reason it is customary in texts at this level to free the real numbers from their decimal shackles and to resort to a qualitative description of their behavior. This will be accomplished here in two steps. In the next section we focus on the arithmetical operations and the notion of order. The subsequent section is devoted to the algebraic analog of the geometrical assumption that the real numbers fill up the number line. An alternate, more constructive approach to the description of the real number system is contained in Edmund Landau's classic book.

Exercises 6.1

1. Prove that there is no rational number r such that $r^2 = 3$.

2. Let p be any prime integer. Prove that there is no rational number r such that $r^2 = p$.

3. Prove that there is no rational number r such that $r^3 = 2$.

4. Let p by any prime integer. Prove that there is no rational number r such that $r^3 = p$.

5. Prove that if p is a prime number and n is any integer greater than 1, then there is no real number r such that $r^n = p$.

6. Assuming that the decimal expansion of the rational number m/n is obtained in the usual way (by the long division process), prove that the decimal expansion of every rational number has the form $N.a_1 \ldots a_s \overline{a_{s+1} \ldots a_t}$ (see Exercise 1.1.8).

7. Explain why the decimal number $.1010010001000010000010\ldots$ is not a rational number.

8. Prove that $\sqrt{6}$ is irrational.

9. Prove that $\sqrt{2} + \sqrt{3}$ is irrational.

10. Prove that $\sqrt{3}/\sqrt{2}$ is irrational.

6.2   ORDERED FIELDS

Before setting out to rigorize some of the arguments presented in the preceding chapters, it is necessary to discuss the properties of the system of real numbers and the features that distinguish it from other number systems.

It is assumed that the readers are familiar with those properties of the integers $\mathbb{Z}$ and the rational numbers (quotients of integers) $\mathbb{Q}$ that are covered in the standard precalculus courses. The letters $\mathbb{N}$, $\mathbb{R}$, denote the non-negative integers and the real numbers respectively. Unless otherwise stated, the letters s, t denote arbitrary integers and the letters k, m, n non-negative integers.

A *set* is a collection of objects called *elements*. If the element a belongs to the set S we write $a \in S$. The *empty* set that contains no elements at all is denoted by $\emptyset$. If S and T are two sets then their *union* $S \cup T$ is the set that consists of all the elements that belong to S or to T or to both. The *intersection* $S \cap T$ of S and T consists of all the elements that belong to both S and T. Thus, if $S = \{1, 2, 5\}$ and $T = \{2, 5, 7, 8\}$, then $S \cup T = \{1, 2, 5, 7, 8\}$ and $S \cap T = \{2, 5\}$. Two sets are said to be *disjoint* if their intersection is empty. If every element of S is also an element of T we say that S is a *subset* of T and write $S \subset T$. Thus, $\{1, 3, 5\} \subset \{1, 2, 3, 4, 5, 6\}$. Given two sets S and T, $S - T$ denotes the set of all the elements of S that do not belong to T. Thus, if $S = \{1, 2, 3, 4, 5, 6\}$ and $T = \{2, 4, 6, 8, 10\}$, then $S - T = \{1, 3, 5\}$ and $T - S = \{8, 10\}$.

A *field* is a set F with two binary operations, denoted by $+$ and $\times$, for which the following hold. For any elements a, b, c of F,

F1. $a + b \in F$, $a \times b \in F$ (*closure*),

F2. $(a + b) + c = a + (b + c)$, $(a \times b) \times c = a \times (b \times c)$ (*associativity*),

F3. $a + b = b + a$, $a \times b = b \times a$ (*commutativity*),

F4. $a \times (b + c) = a \times b + a \times c$ (*distributivity*),

F5. There exist distinct elements $0, 1 \in F$ such that $a + 0 = a$, $a \times 1 = a$ (*identities*),

F6. There exists an element $-a \in F$ such that $a + (-a) = 0$ (*additive inverse*),

F7. If $a \neq 0$ then there exists an element $a^{-1} \in F$ such that $a \times a^{-1} = 1$ (*multiplicative inverse*).

The real numbers with the ordinary addition and multiplication form a field, as do the rational numbers and the complex numbers. Other examples abound (see Exercises 6.2.36, 6.3.4). The integers do *not* form a field since hardly any integers have multiplicative inverses. The set of non-negative real numbers also fails to constitute a field since each of its positive elements fails to have an additive inverse.

Proposition 6.2.1 *If F is a field and $a, b, c \in F$, then*

1. $a + c = b + c$ *implies that $a = b$;*

2. $a \times 0 = 0$;

3. $(-a)b = -(ab)$;

4. $(-a)(-b) = ab$;

5. $ac = bc$ *and $c \neq 0$ imply that $a = b$;*

6. $ab = 0$ *implies that either $a = 0$ or $b = 0$.*

Proof.

1. Since $a + c = b + c$ it follows that $(a + c) + (-c) = (b + c) + (-c)$. Applications of F2, F6, and F5 in succession then yield $a = b$.

2. $0 + a \times 0 = a \times 0 = a \times (0 + 0) = a \times 0 + a \times 0$. Hence, by part 1, $0 = a \times 0$.

3. $ab + (-a)b = b(a + (-a)) = b \times 0 = 0 = ab + (-(ab))$. Hence, by part 1, $(-a)b = -(ab)$.

4. Exercise 38

5. Exercise 39

6. Suppose $b \neq 0$. Then $0 = 0 \times b^{-1} = (ab)b^{-1} = a(bb^{-1}) = a \times 1 = a$. Hence, either $a = 0$ or $b = 0$. $\qquad\square$

A field F is said to be an *ordered field* if there is a binary relation $\leq$ on its elements that has the following properties for any $a, b, c \in F$:

O1. Either $a \leq b$ or $b \leq a$.

O2. If $a \leq b$ and $b \leq a$, then $a = b$.

O3. If $a \leq b$ and $b \leq c$, then $a \leq c$.

O4. If $a \leq b$, then $a + c \leq b + c$.

O5. If $a \leq b$ and $0 \leq c$, then $ac \leq bc$.

Both the systems of real and the rational numbers are ordered. The complex numbers, however, are not, and cannot be, ordered in any manner that is consistent with properties O1 to O5 (Exercise 35). If a and b are distinct elements of F such that $a \leq b$, then we can also write $a < b$. The expression $a \geq b$ is synonymous with $b \leq a$. If n is a positive integer then the power a^n is defined inductively in Exercise 1.

Proposition 6.2.2 *If F is an ordered field and $a, b, c, d \in F$, then*

1. $a \leq b$ *implies* $-b \leq -a$;

2. $a \leq b$ *and* $c \leq d$ *imply that* $a + c \leq b + d$;

3. $a \leq b$ *and* $c \leq 0$ *imply* $bc \leq ac$;

4. $0 \leq a$ *and* $0 \leq b$ *imply that* $0 \leq ab$;

5. $0 \leq a^2$;

6. $0 < 1$;

7. $0 < a$ *implies that* $0 < a^{-1}$;

8. $0 < a < b$ *implies that* $0 < b^{-1} < a^{-1}$;

9. $0 \leq a \leq b$ *and* $0 \leq c < d$ *imply that* $ac < bd$;

10. $0 \leq a < b$ *implies that* $a^n < b^n$ *for every positive integer* n;

11. $1 < a$ *implies that* $a < a^n$ *for every positive integer* n;

12. $-1 \leq a$ *implies that* $(1 + a)^n \geq 1 + na$ *for every positive integer* n.

Proof.

1. Suppose $a \leq b$. Then $-b = a + ((-a) + (-b)) \leq b + ((-a) + (-b)) = -a$.

2. Suppose $a \leq b$ and $c \leq d$. Then $a + c \leq b + c \leq b + d$.

3. Suppose $a \leq b$ and $c \leq 0$. Then $-c \geq 0$ and so $-(ac) = a(-c) \leq b(-c) = -(bc)$. Hence, by 1, $bc \leq ac$.

4. Suppose $0 \leq a$ and $0 \leq b$. Then, by O5, $0 = 0 \times 0 \leq ab$.

5. If $a \geq 0$, this follows from 4. If $a < 0$, then $-a > 0$ and so again $0 \leq (-a)^2 = a^2$.

6. Assume not. Then it follows from O1 that $1 < 0$ and so $-1 > 0$. Hence $1 = (-1)^2 > 0$, contradicting the assumption.

7. Suppose $0 < a$ but $0 \geq a^{-1}$, so that $0 \leq -a^{-1}$. Then $0 = 0 \times 0 \leq a(-a^{-1}) = -1$, contradicting 6.

8. Suppose $0 < a < b$. Then $a^{-1} = b(a^{-1}b^{-1}) > a(a^{-1}b^{-1}) = b^{-1}$.

9. See Exercise 24.

10. See Exercise 25.

11. See Exercise 26.

12. By induction on n. The inequality is clearly valid for $n = 1$, and we now assume its validity for some $n \geq 1$. Then, for $a \geq -1$,

$$(1+a)^{n+1} = (1+a)^n(1+a) \geq (1+na)(1+a) = 1 + na + a + na^2$$
$$\geq 1 + (n+1)a \qquad \Box$$

If a is an element of an ordered field, then a is called a *number* and

$$|a| = a \quad \text{if } a \geq 0 \qquad \text{and} \qquad |a| = -a \quad \text{if } a \leq 0.$$

$|a|$ is called the *absolute value* of a. It is useful to think of $|a|$ as the distance of a from 0, and of $|a - b|$ as the distance between a and b. If $a > 0$ then a is *positive* and if $a < 0$ then a is *negative*.

Proposition 6.2.3 *If F is an ordered field and $a, b \in F$, then*

1. $|a| \geq 0$;

2. $-|a| \leq a \leq |a|$;

3. $-a \leq b \leq a$ *if and only if* $|b| \leq a$;

4. $|ab| = |a| \times |b|$;

1. $|a + b| \leq |a| + |b|$ *(Triangle inequality)*.

Proof.

1, 2. These follow immediately from the definition of $|a|$.

3. If $-a \leq b \leq a$ and $b \geq 0$ then $|b| = b \leq a$. If $-a \leq b \leq a$ and $b < 0$, then $|b| = -b \leq -(-a) = a$. The converse is relegated to Exercise 28.

4. There are four possibilities here, determined by the signs of a and b. In each case the statement follows immediately from the definition of the absolute value.

5. It follows from 2 that $-|a| \leq a \leq |a|$ and $-|b| \leq b \leq |b|$. Consequently, by Proposition 6.2.2.2,

$$-|a| + (-|b|) \leq a + b \leq |a| + |b|$$

or

$$-(|a| + |b|) \leq a + b \leq |a| + |b|.$$

It now follows from 3 that $|a + b| \leq |a| + |b|$.   □

Let F be any field. If P is any property that numbers of F may or may not possess, then

$$\{x \in F \mid x \text{ has property } P\}$$

denotes the set of all the numbers in F that have property P. Thus $\{x \in R \mid x > 0\}$ denotes the set of all the positive real numbers. This notation is used to define a variety of sets. For any two numbers a, b in an ordered field F

$$
\begin{aligned}
(a, b) &= \{x \in F \mid a < x < b\} &&(\textit{open interval}) \\
[a, b] &= \{x \in F \mid a \leq x \leq b\} &&(\textit{closed interval}) \\
[a, b) &= \{x \in F \mid a \leq x < b\} &&(\textit{semiopen interval}) \\
(a, b] &= \{x \in F \mid a < x \leq b\} &&(\textit{semiopen interval}) \\
(a, \infty) &= \{x \in F \mid a < x\} &&(\textit{open ray}) \\
[a, \infty) &= \{x \in F \mid a \leq x\} &&(\textit{closed ray}) \\
(-\infty, b) &= \{x \in F \mid x < b\} &&(\textit{open ray}) \\
(-\infty, b] &= \{x \in F \mid x \leq b\} &&(\textit{closed ray}) \\
(-\infty, \infty) &= F
\end{aligned}
$$

If r is a number, then a *neighborhood* of r is an open interval $(a, b) \subseteq F$ such that $r \in (a, b)$. This neighborhood is centered at r if $r = (a + b)/2$. If ε and a are any numbers, then the ε-*neighborhood of* a is the interval $(a - \varepsilon, a + \varepsilon)$.

If S is a finite set of numbers, then $\max S$ denotes the largest element of S and $\min S$ denotes the smallest element of S.

If S is any set of numbers, and b is any number such that $r \leq b$ for all $r \in S$, then b is said to be an *upper bound* of S and S is said to be *bounded above* by b. Thus, 2 is an upper bound of the set of all the negative real numbers as are 1 and 0. The set of positive numbers, on the other hand, has no upper bound. Similarly, if S is any set of numbers, and b is any number such that $r \geq b$ for all $r \in S$, then b is said to be a *lower bound* of S and S is *bounded below* by b. Thus, -2 is a lower bound of the set of all the positive numbers as are -1 and 0. The set of negative integers, on the other hand, has no lower bound. A set that is bounded both above and below is said to be *bounded*. It is clear that a set S is bounded if and only if there is a number b such that

$$|r| \leq b \qquad \text{for all } r \in S,$$

in which case we say that b *bounds* S. The following easy lemma will prove useful later.

Lemma 6.2.4 *If $S_1, S_2, \ldots, S_n$ are bounded sets, then the set $S_1 \cup S_2 \cup \ldots \cup S_n$ is also bounded.*

Proof. For each $k = 1, 2, \ldots, n$ let b_k be a bound for S_k. Then $b = \max\{b_1, b_2, \ldots, b_n\}$ bounds $S_1 \cup S_2 \cup \ldots \cup S_n$. □

Caveat. Strictly speaking the proofs of this section are incomplete in that they make use of unstated assumptions. Thus the proof of Proposition 6.2.1.1 makes use of the transitivity of equality, the proof of Proposition 6.2.2.1 makes use of the analogous property for inequalities, and the proof of Proposition 6.2.2.12 employs mathematical induction. This is a common practice in books at this level since such concerns, if carried to their extremes, would result in an overly tedious and excessively lengthy exposition.

Exercises 6.2

In Exercises 1–3, F is a field and $a, b \in F$.
1. Define $a^0 = 1$ and $a^{n+1} = a^n$ for $n = 0, 1, 2, \ldots$. Prove that $a^m \times a^n = a^{m+n}$ and $(a^m)^n = a^{mn}$ for $m, n = 0, 1, 2, \ldots$.

2. Prove that $(ab)^n = a^n b^n$ for $n = 0, 1, 2, \ldots$.

3. If $a \neq 0$, and $n > 0$, define $a^{-n} = (a^{-1})^n$. Prove Exercises 1, 2 for all integer values of m, n.

If F is a field, and $a, b \in F$, define $a - b = a + (-b)$. If $b \neq 0$ define $a/b = ab^{-1}$. Suppose that $a, b, c, d, e, f \in F$. Prove that Exercises 4–18 hold in every field. In proving any one of these you may also rely on any of the previous ones.
4. $(a + b)(c + d) = ac + ad + bc + bd$

5. $a(b - c) = ab - ac$

6. $(a + b)(a - b) = a^2 - b^2$

7. $(a + b)^2 = a^2 + 2ab + b^2$

8. $(a - b)^2 = a^2 - 2ab + b^2$

9. $(a + b)^3 = a^3 + 3a^2 b + 3ab^2 + b^3$

10. $(a - b)^3 = a^3 - 3a^2 b + 3ab^2 - b^3$

11. $1 + a + a^2 + \ldots + a^n = (1 - a^{n+1})/(1 - a)$ if $a \neq 1$.

12. If $a + x = a$, then $x = 0$.

13. If $ax = a$ and $a \neq 0$, then $x = 1$.

14. If $a + x = 0$, then $x = -a$.

15. If $ax = 1$, then $x = a^{-1}$.

16. If $ax + b = 0$ and $a \neq 0$, then $x = -(b/a)$.

17. If $ax^2 + bx + c = 0$ and $a \neq 0$, then

$$\left(x + \frac{b}{2a}\right)^2 = \frac{b^2 - 4ac}{4a^2}.$$

18. There exists a unique pair of numbers x, y such that $ax + by = c$ and $dx + ey = f$ if and only if $ae - bd \neq 0$.

Let $a, b, c, d, e, f \in F$ and suppose $a/b = c/d = e/f$. Prove Exercises 19–22.

19. $\dfrac{a + b}{a - b} = \dfrac{c + d}{c - d}$

20. $\dfrac{a}{b} = \dfrac{a + c + e}{b + d + f}$

21. $\dfrac{a}{b} = \dfrac{a - 3c + 2e}{b - 3d + 2f}$

22. $\dfrac{a^3}{b^3} = \dfrac{a^3 + 9c^3 - 15e^3}{b^3 + 9d^3 - 15f^3}$

Let $a, b, c, d, e, f \in F$ where F is an ordered field. Prove Exercises 23–33.

23. If $a < 2^{-1}$ then $(1 - a)^{-1} < 2$.

24. Proposition 6.2.2.9.

25. Proposition 6.2.2.10.

26. Proposition 6.2.2.11.

27. If $0 \leq a, b$ and $a^n \leq b^n$ for some positive integer n then $a \leq b$.

28. Proposition 6.2.3.3 (i.e., if $|b| \leq a$ then $-a \leq b \leq a$).

29. If $a, b, c, d > 0$ and $a/b < c/d$ then

$$\frac{a}{b} < \frac{a + c}{b + d} < \frac{c}{d}.$$

30. $(a^2 + b^2)/2 \geq ab$

31. $(a^2 + b^2)(c^2 + d^2) \geq (ac + bd)^2$.

32. $(a^2 + b^2 + c^2)(d^2 + e^2 + f^2) \geq (ad + be + cf)^2$.

33. $(a^2 + b^2)(b^2 + c^2)(c^2 + a^2) \geq 8a^2b^2c^2$.

34. Prove that if $a, b \in F$ where F is an ordered field, then
   a) $|a + b| \geq |a| - |b|$
   b) $|a - b| \geq |a| - |b|$
   c) $||a| - |b|| \leq |a - b|$ (Hint: Use Propositions 6.2.3.2 and 6.2.3.5.)

35. Prove that there is no binary relation $\leq$ which turns the complex numbers into an ordered field.

36. Let $\mathbb{Q}[\sqrt{2}]$ denote the set of all numbers of the form $a + b\sqrt{2}$ where a and b are rational numbers. Prove that this set of numbers forms an ordered field with respect to ordinary addition and multiplication.

37. The field $\mathbb{Z}_2$ contains two elements 0 and 1 only and its addition and multiplication are defined as $0 + 0 = 1 + 1 = 0$, $0 + 1 = 1 + 0 = 1$, $0 \times 0 = 0 \times 1 = 1 \times 0 = 0$, $1 \times 1 = 1$. Prove that $\mathbb{Z}_2$ is a field. Prove that there is no binary relation $\leq$ that turns $\mathbb{Z}_2$ into an ordered field.

38. Proposition 6.2.1.4.

39. Proposition 6.2.1.5.

40. Suppose $S = \{1, 2, 3, 4, 5, 6\}$, $T = \{0, 2, 4, 6, 8\}$, $U = \{0, 1, 3, 5, 7, 9\}$. Find the following:

   a) $S \cup T$ b) $S \cup U$ c) $T \cup U$ d) $S \cap T$
   e) $S \cap U$ f) $T \cap U$ g) $S - T$ h) $T - S$
   i) $S - U$ j) $U - S$ k) $T - U$ l) $\max S$
   m) $\min S$ n) $\max T$ o) $\min T$

41. Prove that if x is a real number and n is a positive integer, then there is another integer m such that $|x - (m/n)| < 1/n$.

42. For every real number x, let $[x]$ denote the largest integer not exceeding x. Prove that if x is an irrational number and n is a positive integer, then there are positive integers $b < n$ and a such that $|bx - a| < 1/n$.

6.3   COMPLETENESS AND IRRATIONAL NUMBERS

All the properties listed in Section 6.2 as either definitions or propositions hold for both the rational and the real number systems. By pointing out a qualitative difference between them, it is now demonstrated that these two number systems are indeed distinct.

If b^* is an upper bound for S such that $b^* \leq b$ for every upper bound b of S, then b^* is said to be the *least upper bound* of S (see Exercise 2 regarding uniqueness). For example, if S is the set of all the negative numbers, then every non-negative number is an upper bound of S, and 0 is the least upper bound of S. The set of all positive numbers has no upper bound at all, and the closed interval $[1, 2]$ has 2 as its least upper bound.

The number d^* is the *greatest lower bound* of S provided d^* is a lower bound of S and $d \leq d^*$ for all lower bounds d of S. Every negative number is a lower bound of the set of positive numbers, and 0 is the greatest lower bound of this set. Every number less than -2.3 is a lower bound of the interval $[-2.3, 5)$ and -2.3 is the greatest lower bound of this interval.

The empty set $\emptyset$ exhibits a curious behavior with regard to these concepts. The way these definitions are worded, the empty set $\emptyset$ is bounded above by every number b since the requirement $r \leq b$ is vacuously satisfied (there are no elements in $\emptyset$ to violate it). However, since every number is an upper bound of $\emptyset$, and there is no least number, it follows that the set $\emptyset$ has no least upper bound. We now ask whether there are any non-trivial, that is, non-empty sets that also exhibit this anomaly. The answer depends on the ordered field in question.

For the sake of the ensuing discussion the reader is asked to temporarily accept the existence of a real number $\sqrt{2}$ such that $(\sqrt{2})^2 = 2$, a claim that will be justified below as well as again in Section 11.4. Of course, it is already known that this number, if it exists, is not rational.

Looking at the real numbers first, let $S = \{r \in \mathbb{R} \mid r^2 \leq 2\}$, that is, the set of all real numbers whose square is at most 2. It is clear that this set is in fact the interval $[-\sqrt{2}, \sqrt{2}]$ and so it is bounded and has $\sqrt{2}$ as its least upper bound. Next let us examine the anologous set within the context of the rational numbers $\mathbb{Q}$. If $T = \{r \in \mathbb{Q} \mid r^2 \leq 2\}$, *does T have a <u>rational</u> upper bound and does it have a <u>rational</u> least upper bound?* Some work with the calculator indicates that T contains the rational numbers 1, 1.4, 1.41, 1.414, 1.4142, 1.41421, etc., and that each of the numbers 1.5, 1.42, 1.415, 1.4143, 1.41422 etc. is a rational upper bound for T. The least upper bound is more elusive. By rights, $\sqrt{2}$ should again have been the least upper bound of T. However, by Proposition 6.1.1, there is no rational number whose square is 2. Hence, just like the empty set $\emptyset$, the bounded set T does not have a least upper bound in the context of the rational numbers.

This difference between the real and the rational number systems turns out to be the key qualitative difference between them. The real number system satisfies the Completeness Axiom below whereas the rational number system does not.

Axiom 6.3.1 (Completeness Axiom) *Every non-empty set of real numbers that is bounded above has a least upper bound.*

The *real number system* is now formally defined as an ordered field that also satsifies the Completeness Axiom. It so happens that any ordered field that satisfies the completeness axiom is indistinguishable from the real number system in a formal sense whose explanation falls outside the scope of this text. The significance of this axiom was first recognized by Bernard Bolzano (1781–

1848) and his contributions on this topic vis-a-vis continuity will be discussed in greater detail in Chapter 11. Some of the important implications that the Completeness Axiom has for the convergence of series will be discussed in Chapters 8 and 9. Other consequences will be discussed momentarily, after the use of the word completeness to describe this axiom is justified. Informally speaking, an ordered field that contains the rationals <u>and</u> satisfies the completeness axiom necessarily "fills out" the number line. To see this, let X be any point of the number line and let (see Fig. 6.2)

$$S = \{r \in \mathbb{Q} \mid P_r \text{ lies to the left of } X \text{ on the number line}\}.$$

Then S is a non-empty set of rational numbers that is bounded above by every rational number r' such that $P_{r'}$ lies to the right of X. The completeness axiom asserts the existence of a least upper bound b of S and this b has the property that $P_b = X$. This argument must of necessity remain informal since a rigorous proof would require a formal definition and study of the straight line, a non-trivial task that lies outside the scope of this text.

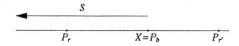

Fig. 6.2 The real line is complete.

The Completeness Axiom is used now to prove rigorously several facts of interest. The first of these is logically equivalent to the property promulgated by Archimedes in his introductory comments in Appendix A. It also excludes the transfinite numbers from the realm of real numbers.

Proposition 6.3.2 *Given any real number r, there exists an integer s such that $s \leq r < s + 1$.*

Proof. We first consider the case $r \geq 0$. Let $S = \{n \in \mathbb{N} \mid n \leq r\}$. This set is bounded above by r and is not empty since $0 \in S$. Let b^* be the least upper bound of S. Then $b^* - 1$ is not a bound of S and so there is an integer s in S such that $b^* - 1 < s \leq r$ (see Fig. 6.3). Clearly, $b^* < s + 1$ and so $s + 1$ is an integer <u>not</u> in S, i.e., $s + 1 > r$. Thus, $s \leq r < s + 1$.

If $r < 0$, and r is not an integer, then by the first part of the proof there exists an integer t such that $t < -r < t + 1$. Hence, $-t - 1 < r < -t$, and so we may choose $s = -t - 1$.

Finally, if r is a negative integer, then we can choose $s = r$. □

Fig. 6.3 Bounding a real number by integers.

As another interesting application of the Completeness Axiom we prove the existence of roots. Let n be a positive integer and let r be a non-negative real number. If s is a non-negative real number such that $s^n = r$, then we say that s is the nth *root* of r and write $s = \sqrt[n]{r}$ with the understanding that $\sqrt[2]{r} = \sqrt{r}$. It is now proved that these roots always exist in $\mathbb{R}$. This proof is considerably more intricate than that of the previous proposition and, since an alternative and much easier proof is offered in Corollary 11.4.3, this one might well be omitted it on first reading.

Lemma 6.3.3 *Let n be a positive integer and r a real number. Then*

$$(1+r)^n < 1 + 2^n r \qquad \text{if } 0 < r \leq 1.$$

Proof. It follows from two applications of the finite geometric progression (Proposition 1.1.2 and Exercise 6.2.11) that for $0 < r < 1$

$$(1+r)^n - 1 = [(1+r) - 1] \times [(1+r)^{n-1} + (1+r)^{n-2} + \cdots + (1+r) + 1]$$
$$< r(2^{n-1} + 2^{n-2} + \cdots + 2 + 1) = r(2^n - 1) < 2^n r.$$

The desired inequality now follows immediately. $\qquad\qquad\square$

Proposition 6.3.4 *For every non-negative real number r and for every positive integer n, $\sqrt[n]{r}$ exists (in $\mathbb{R}$).*

Proof. Let $S = \{x \in \mathbb{R} \mid x^n \leq r\}$. The set S contains 0 and if $x \in S$ then, by Proposition 6.2.2

$$x^n \leq r < 1 + r < (1+r)^n$$

so that, by Exercise 6.2.27, $x < 1 + r$. Thus S is bounded above and so, by the Completeness Axiom, S has a least upper bound, say b. We now show by contradiction that $b^n = r$. Suppose $b^n \neq r$.

If $b^n < r$ then for every integer $m \geq 2^n b^n / (r - b^n)$ we have by Lemma 6.3.3

$$\left(b + \frac{b}{m}\right)^n = b^n \left(1 + \frac{1}{m}\right)^n < b^n \left(1 + \frac{2^n}{m}\right) \leq b^n \left(1 + \frac{r - b^n}{b^n}\right) = r.$$

This means that $b + (b/m) \in S$, contradicting the fact that b is an upper bound of S.

If $b^n > r$, then for every positive integer m, $b - (b/m)$ is not an upper bound of S and hence there exists an $x \in S$ such that $x > b - (b/m)$. If we now choose an integer $m \geq nb^n / (b^n - r)$, then by Lemma 6.2.2.12

$$r \geq x^n > \left(b - \frac{b}{m}\right)^n = b^n \left(1 - \frac{1}{m}\right)^n \geq b^n \left(1 - \frac{n}{m}\right)$$
$$\geq b^n \left(1 - \frac{b^n - r}{b^n}\right) = r,$$

which is impossible. Hence, $r = b^n$, or, $b = \sqrt[n]{r}$.   □

It was Newton who proposed the definitions $a^{m/n} = \sqrt[n]{a^m}$ and $a^{-r} = 1/a^r$ for every positive real number a, positive rational number r, and positive integers m and n (see Appendix C). The resultant Fractional Binomial Theorem shows this to be much more than a mere notational convenience.

Every real number that is not a rational number is said to be *irrational*. Exercise 6.1.5 asserts that if p is a prime and $n > 1$ is an integer, then $\sqrt[n]{p}$ is irrational. Each of the several parts of the following proposition is proved by a straightforward reductio ad absurdum (Exercise 6).

Proposition 6.3.5 *If r is a non-zero rational number and a is an irrational number then the numbers $a \pm r$, ar, a/r, r/a are all irrational.*

The following proposition says that both rational and irrational numbers can be found everywhere on the real line. It will later facilitate the construction of some interesting functions.

Proposition 6.3.6 *If $a < b$ are real numbers, then there exist both a rational number and an irrational number between a and b.*

Proof. Let t be any integer greater than $2\sqrt{2}/(b-a)$ so that

$$\frac{1}{t} < \frac{\sqrt{2}}{t} < \frac{b-a}{2}.$$

Let s be an integer such that $s \le ta < s + 1$. It then follows that

$$a < \frac{s}{t} + \frac{1}{t} < \frac{s}{t} + \frac{\sqrt{2}}{t} < a + \frac{b-a}{2} = \frac{a+b}{2} < b.$$

Thus, $(s+1)/t$ and $(s+\sqrt{2})/t$ are the required rational and irrational numbers.
     □

Exercises 6.3

1. Determine the least upper bound and the greatest lower bound, if any, of each of the following sets of real numbers. In each case specify whether the said bound does or does not belong to the given set.

a) $\left\{ 1, \frac{1}{2}, \frac{1}{3}, \dots \right\}$ b) $(1, 10)$ c) $[-1, 10)$

d) $\left\{ \frac{(-1)^n}{n} \,\middle|\, n = 1, 2, 3, \dots \right\}$ e) $(-\infty, \sqrt{2})$ f) $(-\infty, -\sqrt{3}]$

g) The positive reals   h) The negative reals
i) The non-positive reals   j) The non-negative reals
k) The positive rationals   l) The negative rationals
m) The non-positive rationals   n) The non-negative rationals
o) $(-\sqrt{3}, \sqrt[3]{17})$ p) $\{1, \sqrt{17}, -2, 4\}$
q) $\{.1, .11, .111, .1111, \dots\}$ r) $\{.1, .01, .001, .0001, \dots\}$

s) $\left\{ 1 - \dfrac{1}{n} \,\middle|\, n = 1, 2, 3, \dots \right\}$ t) $\left\{ 1 + \dfrac{1}{n} \,\middle|\, n = 1, 2, 3, \dots \right\}$

u) $\left\{ 1 + \dfrac{(-1)^n}{n} \,\middle|\, n = 1, 2, 3, \dots \right\}$ v) $\left\{ (-1)^n \left(1 - \dfrac{1}{n} \right) \,\middle|\, n = 1, 2, 3, \dots \right\}$

2. Prove that every non-empty bounded set has exactly one greatest lower bound.

3. Let $a, b \in \mathbb{R}$. Prove that if $a, b \geq 0$, then $\sqrt{ab} \leq (a + b)/2$.

4. For any prime number p, let $\mathbb{Q}[\sqrt{p}]$ denote the set of all numbers of the form $a + b\sqrt{p}$ where a and b are rational numbers. Prove that this set of numbers forms an ordered field with respect to ordinary addition and multiplication.

5. Prove that $\sqrt[n]{r}$ is unique.

6. Prove Proposition 6.3.5.

7. Let M be a subset of $\mathbb{R}$ such that $M \neq \mathbb{R}$ and $M \supset (-\infty, u)$ for some u. Prove that there exists a number U such that
   a) $M \supset (-\infty, U)$
   b) if $M \supset (-\infty, r)$ then $r \leq U$.

8. Suppose $a \neq 0$ and b is any real number. Prove that $\{an + b \mid n = 0, 1, 2, \dots\}$ is unbounded.

6.4* THE EUCLIDEAN PROCESS

The fascination with irrational numbers goes back to the beginnings of pure mathematics. In his *Elements* Euclid offered a characterization of irrational numbers that bypasses the arithmetical lemmas needed to validate the argument of Proposition 6.1.1. His approach was based on the process of long division.

Proposition 6.4.1 (Long division) *Given any two real numbers $a > 0$ and b, there exist a unique integer q and a unique real number $0 \leq r < a$ such that*

$$b = qa + r \tag{1}$$

Proof. By Proposition 6.3.2 there exists an integer q such that

$$q \leq \frac{b}{a} < q + 1 \quad \text{or} \quad qa \leq b < qa + a.$$

To obtain Equation (1) it only remains to set $r = b - qa$.

Suppose q' is another integer and $0 \leq r' < a$ is another real number such that $b = q'a + r'$. Then $qa + r = q'a + r'$ and so

$$(q - q')a = (r' - r).$$

However, $|r' - r| < a$ and so, since $q - q'$ is an integer, we must have $q - q' = 0$, or $q = q'$. It now follows immediately that $r' = r$. Hence the uniqueness of q and r. □

The number r in the statement of Proposition 6.4.1 is called the *remainder* when b is *divided by* a. The Euclidean process is based on a function E that assigns to every pair (a, b) of real numbers, $0 < a \leq b$ the pair $E(a, b) = (r, a)$ where r is the remainder of b when divided by a. Thus,

$$E(10, 73) = (3, 10)$$
$$E(3, 10) = (1, 3)$$
$$E(1, 3) = (0, 1)$$

and $E(0, 1)$ is undefined. Similarly,

$$E\left(\frac{1}{3}, \frac{5}{2}\right) = \left(\frac{1}{6}, \frac{1}{3}\right)$$
$$E\left(\frac{1}{6}, \frac{1}{3}\right) = \left(0, \frac{1}{6}\right)$$
$$E(1, \sqrt{2}) = (\sqrt{2} - 1, 1)$$
$$E(\sqrt{2} - 1, 1) = (1 - 2(\sqrt{2} - 1), \sqrt{2} - 1) = (3 - 2\sqrt{2}, \sqrt{2} - 1).$$

The Euclidean process assigns to every pair of real numbers (a, b) such that $0 < a \leq b$ the (possibly finite) list

$$\mathcal{E}(a, b) = \{E(a, b), E^2(a, b), E^3(a, b), \dots\}$$

where $E^1(a, b) = E(a, b)$ and $E^{n+1}(a, b) = E(E^n(a, b))$ for $n = 1, 2, 3, \dots$. For example,

$$\mathcal{E}(10, 73) = \{(3, 10), (1, 3), (0, 1)\}$$
$$\mathcal{E}\left(\frac{1}{3}, \frac{5}{2}\right) = \left\{\left(\frac{1}{6}, \frac{1}{3}\right), \left(0, \frac{1}{6}\right)\right\}.$$

It is clear from the definition of $E(a, b)$ that in general the list $\mathcal{E}(a, b)$ has the format

$$\mathcal{E}(a, b) = \{(a_1, a), (a_2, a_1), (a_3, a_2), (a_4, a_3), \dots\} \tag{2}$$

where
$$a > a_1 > a_2 > a_3 > \ldots \geq 0.$$

For every pair (a, b), let $\lambda(a, b)$ be the largest integer for which $E^\lambda(a, b)$ is defined. Thus, $\lambda(10, 73) = 3$, $\lambda\left(\frac{1}{3}, \frac{5}{2}\right) = 2$. It turns out that $\lambda(1, \sqrt{2})$ does not exist, a fact that will be demonstrated later. In such cases we write $\lambda(a, b) = \infty$. Clearly $\lambda(a, b)$ equals the length of the list $\mathcal{E}(a, b)$ of Equation (2).

Lemma 6.4.2 *If $0 < s \leq t$ are integers, then $\lambda(s, t)$ is finite.*

Proof. Since s and t are integers, then it follows from Proposition 6.4.1 by a straightforward induction argument that all the $a_i's$ of Equation (2) are non-negative integers. Thus there must exist an integer n such that $a_n = 0$ and so $\lambda(s, t) = n$. $\square$

Lemma 6.4.3 *If $0 < a \leq b$, $c > 0$, and $E(a, b) = (r, a)$, then*

1. *$E(ac, bc) = (rc, ac)$;*

2. *$\lambda(ac, bc) = \lambda(a, b)$.*

Proof. Since $b = qa + r$ with $0 \leq r < a$ if follows that $bc = q(ac) + (rc)$ with $0 \leq rc < ac$. It follows from the uniqueness guaranteed by Proposition 6.4.1 that $E(ac, bc) = (rc, ac)$. Implicit in this is the observation that $E(ac, bc)$ is defined if and only if $E(a, b)$ is defined. A straightforward induction on n shows that $E^n(ac, bc)$ exists if and only if $E^n(a, b)$ exists for $n = 1, 2, 3, \ldots$. Thus, $\lambda(ac, bc) = \lambda(a, b)$.

The following restates the second proposition of Book X of Euclid's *Elements*.

Proposition 6.4.4 *If $0 < a \leq b$ are real numbers such that a/b is a rational number then $\lambda(a, b)$, is finite.*

Proof. If a/b is a rational number, then $a/b = s/t$ for some non-negative integers s, t. Let $c = a/s = b/t$. Then, by Lemma 6.4.3 $\lambda(a, b) = \lambda(cs, ct) = \lambda(s, t)$ which, by Lemma 6.4.2, is finite. $\square$

Proposition 6.4.5 *The number $\sqrt{2}$ is irrational.*

Proof. It will be proven that $\sqrt{2} + 1$ is irrational and the desired conclusion then follows immediately. This irrationality is established by proving, by contradiction, that $\lambda(1, \sqrt{2} + 1)$ is infinite. Suppose, to the contrary, that $\lambda(1, \sqrt{2} + 1)$ is finite. Then

$$E(1, \sqrt{2} + 1) = (\sqrt{2} - 1, 1) = (1 \times (\sqrt{2} - 1), (\sqrt{2} + 1) \times (\sqrt{2} - 1))$$
$$= (1 \times c, (\sqrt{2} + 1) \times c) \qquad \text{where } c = \sqrt{2} - 1.$$

Hence,

$$\lambda(1, \sqrt{2} + 1) = 1 + \lambda(E(1, \sqrt{2} + 1)) = 1 + \lambda(1 \times c, (\sqrt{2} + 1) \times c)$$
$$= 1 + \lambda(1, \sqrt{2} + 1)$$

which is impossible. Thus $\lambda(1, \sqrt{2} + 1)$ is infinite and so $(\sqrt{2} + 1)/1 = \sqrt{2} + 1$ is irrational. $\qquad\square$

There is another very important application of the Euclidean process, but since it is not germane to the subsequent material, its proof is relegated to Exercise 10.

Proposition 6.4.6 *Suppose $0 < s \leq t$ are integers. If $\lambda = \lambda(s, t)$ then $E^\lambda(s, t) = (0, g)$, where g is the greatest common divisor of s and t.*

Example 6.4.7 If $a = 216$ and $b = 444$, then

$$\mathcal{E}(a, b) = \{(12, 216), (0, 12)\}$$

and so 12 is the greatest common divisor of 216 and 444.

Exercises 6.4

1. Compute $\mathcal{E}(a, b)$ for $a = 37$, $b = 1000$.

2. Compute $\mathcal{E}(a, b)$ for $a = 144$, $b = 243$.

3. Compute $\mathcal{E}(a, b)$ for $a = 1/2$, $b = 4/3$.

4. Compute $\mathcal{E}(a, b)$ for $a = 2/3$, $b = 3/4$.

5. Compute $E(1, \sqrt{3})$ and $E^2(1, \sqrt{3})$.

6. Compute $E(1, \sqrt{6})$ and $E^2(1, \sqrt{6})$.

   Exercises 7–9 are to be solved on the basis of the Euclidean process.
7. Prove that $\sqrt{5}$ is irrational. (Hint: Work with $\sqrt{5} + 2$.)

8. Prove that if n is any integer, then $\sqrt{n^2 + 1}$ is irrational.

9. Prove that if $0 < a \leq b$ and a/b is irrational, then $\lambda(a, b)$ is infinite.

10. Prove Proposition 6.4.6.

11. Find the greatest common divisor of 234 and 432.

12. Find the greatest common divisor of 3,652 and 51,414.

13. Find the greatest common divisor of 123,456 and 234,561.

6.5 FUNCTIONS

During the fifteenth and sixteenth centuries mathematicians found that the representation of such concepts as *the given quantity* or *the required quantity* by letters such as a or x added to the clarity of both their exposition and thinking. Eventually these quantities were divided into *known* or *constant quantities*, commonly denoted by $a, b, c \ldots$ and *unknown* or *variable quantities* for which Descartes introduced the symbols x, y, z. Functions, on the other hand, were never explicitly mentioned, although they were implicit in many discussions. Instead, mathematicians simply assumed that some of the variables were related by equations such as $x^3 - 3x^2y^3 - y + 7 = 0$ from which one could, if necessary, extract y and express it in terms of x, as was illustrated in Section 4.2.

The word *function* first appeared in a manuscript of Leibniz written in 1673 as a general term for various geometric quantities associated with a variable point on the curve. Johann Bernoulli (1667–1748), who was in communication with Leibniz on this and many other topics, was the first to explicitly define the word function in 1718 in a manner that resembles its usage today:

> One calls here a function of a variable a quantity composed in any manner whatever of this variable and constants.

The composition of the variable and the constants could employ the four arithmetical operations, algebraic operations such as taking roots, or the so called *transcendental* operations such as taking logarithms of numbers and sines of angles. It was common to paraphrase this definition by saying that the function $f(x)$ depended *analytically* on the variable x. While there were some disagreements on the exact nature of this dependence, this view of functions held sway for a hundred years. One of the first to explicitly abandon the analytic dependence of $f(x)$ on x was Fourier who wrote in 1822:

> In general the function $f(x)$ represents a succession of values or ordinates each of which is arbitrary. An infinity of values being given to the abscissa x, there are an equal number of ordinates $f(x)$. All have actual numerical values, either positive or negative or null. We do not suppose these ordinates to be subject to a common law; they succeed each other in any manner whatever, and each of them is given as it were a single quantity.

Peter Lejeune Dirichlet's (1805–1854) definition of 1837 is even closer to the modern version:

> Let us suppose that a and b are two definite values and x is a variable quantity which is to assume, gradually, all values located between a and b. Now, if to each x there corresponds a unique, finite $y \ldots$, then y is called a $\ldots$ function of x for this interval. It is, moreover, not at all necessary, that y depends on x in this whole interval according to the same law; indeed, it is not necessary to think of only relations that can be expressed by mathematical operations.

Every function in the sense of Bernoulli is also a function in the sense of Fourier and Dirichlet. However, the latter allowed for functions that are excluded by Bernoulli's definition. Such is the case for the *Dirichlet function*

$$D(x) = \begin{cases} 1 & \text{if } x \text{ is rational and } 0 \le x \le 1 \\ 0 & \text{if } x \text{ is irrational and } 0 \le x \le 1 \end{cases}$$

This function is well defined in the sense of Fourier and Dirichlet because the value of x completely determines the value of $f(x)$, albeit the determination is verbal and not mathematical (or analytical). However, because $D(x)$ is described in terms they would have considered as non-analytical, the mathematicians of the eighteenth century would not have considered it as a function. Interestingly, it was considerations in applied mathematics that caused mathematicians to abandon the Bernoulli definition. Specifically, it was in the context of their work on the vibrating string that Jean Le Rond D'Alembert (1717–1783), Euler, and others had to face up to the fact that a natural initial position for the string has the shape of Figure 6.4 which is not the graph of

Fig. 6.4   A pulled string.

any expression that they would have considered to be analytical. D'Alembert took a conservative stance and excluded this angular shape from his considerations, thereby considerably weakening his ground-breaking work on sound. Euler, on the other hand, was willing to relax the analyticity requirement and to allow for shapes that, in modern parlance, are only piecewise analytic. Fourier and Dirichlet's further relaxation of the definition of functions was in all likelihood motivated by their work on trigonometric series. Certainly they were aware that the limit of some trigonometric series had discontinuous graphs and would have found it awkward to say that some convergent series converged to a function whereas others converged to a non-function. By the year 1867 Riemann found uses for functions that were every bit as strange looking as Dirichlet's $D(x)$. Such are the considerations that explain our decision to adopt the seemingly vague definition below, which, it might be added, is still not the most formal of the current definitions.

Let D be a set of real numbers. A *function* $f : D \to \mathbb{R}$ is a rule that assigns a number $f(x)$ to every element x of D. For example, the rule $f(x) = \sqrt{1 - x^2}$ defines a function $f : [-1, 1] \to \mathbb{R}$. On the other hand, the rule $f(x) = (x^2 - 1)^{-1/2}$ defines a function $f : D \to \mathbb{R}$, where $D = (-\infty, -1) \cup (1, \infty)$. The set D is called the *domain* of the function $f : D \to \mathbb{R}$ and, in this text, consists of an interval or a ray or the union of some intervals and/or rays. The domain will quite frequently be left implicit, unless its nature turns out to be significant. Correspondingly, the notation $f : D \to \mathbb{R}$ will usually be abbreviated to f.

The four arithmetical operations on the real numbers induce analogous operations on functions. Thus, given two functions $f : D \to \mathbb{R}$ and $g : E \to \mathbb{R}$, the functions $f + g$, $f - g$, $f \times g$, f/g are defined by the following rules:

$$(f + g)(x) = f(x) + g(x) \qquad x \in D \cap E$$
$$(f - g)(x) = f(x) - g(x) \qquad x \in D \cap E$$
$$(f \times g)(x) = f(x)g(x) \qquad x \in D \cap E$$
$$\left(\frac{f}{g}\right)(x) = \frac{f(x)}{g(x)} \qquad x \in D \cap E \text{ and } g(x) \neq 0.$$

Given a function $f : D \to \mathbb{R}$, the set $\{f(x) \mid x \in D\}$ is called the *range* of f and is denoted by $f(D)$. Thus, the range of the function determined by $f(x) = \sqrt{1 - x^2}$ is $[0, 1]$ whereas the range of the function determined by $f(x) = (x^2 - 1)^{-1/2}$ is $(0, \infty)$. Given two functions $f : D \to \mathbb{R}$ and $g : E \to \mathbb{R}$, their *composition* $f \circ g$ is defined by the rule

$$(f \circ g)(x) = f(g(x)) \qquad \text{whenever } x \in E \text{ and } g(x) \in D.$$

Two functions $f, g : D \to \mathbb{R}$ are said to be *equal* if $f(x) = g(x)$ for all $x \in D$. A function $f : D \to \mathbb{R}$ is said to be *one-to-one* if $f(x_1) \neq f(x_2)$ whenever $x_1 \neq x_2$. Accordingly, the function $f(x) = x^3$ is one-to-one whereas the function $g(x) = x^2$ is not one-to-one as $g(3) = 9 = g(-3)$. The functions $f : D \to \mathbb{R}$ and $g : E \to \mathbb{R}$ are said to be *inverses* of each other provided that $(f \circ g)(x) = x$ for all $x \in E$ and $(g \circ f)(x) = x$ for all $x \in D$. For example, if D is the set of all positive real numbers, then the functions $f(x) = \sqrt{x}$ and $g(x) = x^2$ are inverses of each other. A function $f : D \to \mathbb{R}$ is said to be *increasing, decreasing, strictly increasing,* or *strictly decreasing* if for every $x_1, x_2 \in D$ such that $x_1 < x_2$ we have $f(x_1) \leq f(x_2)$, $f(x_1) \geq f(x_2)$, $f(x_1) < f(x_2)$, $f(x_1) > f(x_2)$, respectively.

Exercises 6.5

1. Show that if $f : \mathbb{Q} \to \mathbb{R}$ has the properties that $f(0) = 0$ and $f(x + y) = f(x) + f(y)$ for all $x, y \in \mathbb{Q}$, then there is a number c such that $f(x) = cx$ for all $x \in \mathbb{Q}$.

2. Show that if $f : \mathbb{Q} \to \mathbb{R}$ has the properties that $f(0) = 1$ and $f(x + y) = f(x) \times f(y)$ for all $x, y \in \mathbb{Q}$, then there is a number c such that $f(x) = c^x$ for all $x \in \mathbb{Q}$.

In Exercises 3–6 the domain of each of the functions f, g, h is $\mathbb{R}$.

3. Prove that $(f \circ g) \circ h = f \circ (g \circ h)$.

4. Prove that $(f + g) \circ h = (f \circ h) + (g \circ h)$.

5. Is it true that $f \circ (g + h) = (f \circ g) + (f \circ h)$?

6. Prove that if f and g are one-to-one functions, then so is $f \circ g$ one-to-one.

7. Prove that if $f : D \rightarrow \mathbb{R}$ is either strictly increasing or strictly decreasing then it is one-to-one.

8. Is an increasing function necessarily one-to-one?

9. Describe the main mathematical achievements of
   a) Johann Bernoulli
   b) Jean Le Rond d'Alembert
   c) Peter Lejeune Dirichlet
   d) Bernhard Riemann

10. Use your own words to describe the difference between Johann Bernoulli's concept of a function and those of Fourier and Dirichlet.

Chapter Summary

Following an informal discussion of irrational numbers, the real number system was defined by means of a list of axioms. These are of three types: the field axioms that relate to the standard arithmetical operations, the order axioms that deal with the relative sizes of numbers, and the completeness axiom that can be viewed as identifying the real numbers with the points of the straight line. An optional section also explained Euclid's treatment of the irrational numbers. The last section is devoted to both the definition and the history of the concept of a function.

7

Sequences and Their Limits

The notion of convergence is formalized in terms of sequences and limits. Convergence is formulated in both a geometric and an algebraic manner. Several specific and useful limits are computed, and some theorems and propositions that facilitate the computation of limits are proved.

7.1 THE DEFINITIONS

Arguably, the most important concepts developed by mathematicians in their long and arduous struggle to rigorize calculus were those of a *sequence* and its *limit*. These turned out to serve a dual purpose. They constitute the means by which the notion of the convergence of an infinite series can be defined and, less obviously, they also make possible rigorous definitions of such concepts as continuity and differentiability. The modern and formal treatment of limits of sequences is credited to Augustin-Louis Cauchy (1789–1857) whose definition of convergence is paraphrased as Theorem 7.1.4 below.

A *sequence* is an infinite list of real numbers. The best known sequence is that of the *natural numbers* $0, 1, 2, 3, 4, \ldots$. Other examples of sequences are

$$1, \frac{1}{\sqrt{2}}, \frac{1}{\sqrt{3}}, \frac{1}{\sqrt{4}}, \ldots \tag{1}$$

$$3, 3, 3, 3, \ldots \tag{2}$$

$$1, -1, 1, -1, \ldots \tag{3}$$

$$1, 2, 4, 8, \ldots \tag{4}$$

Somewhat more formally, the sequence $\{a_n\}$ can be defined as a function $f : \mathbb{N} \to \mathbb{R}$ where $f(n) = a_n$ for $n = 0, 1, 2, \ldots$. Thus, the sequence of natural numbers can be thought of as the function $f(n) = n$, whereas the sequences (1–4) can be identified with the functions

$$f(n) = \frac{1}{\sqrt{n+1}},$$
$$f(n) = 3,$$
$$f(n) = (-1)^n,$$
$$f(n) = 2^n,$$

respectively. It should be stressed that while this formal definition has some advantages in more advanced contexts, that will not be the case in this text. It is customary to denote the sequence $a_0, a_1, a_2, \ldots$ by the shorthand notation

$$\{a_n\}_{n=0}^{\infty} \quad \text{or simply} \quad \{a_n\}.$$

The presence of the index n will serve to distinguish the sequence $\{a_n\}$ from the set $\{a\}$. If it so happens that a_0 is undefined, as is the case for the sequence $\{1/n\}$, it will be ignored. Thus, it will be understood that

$$\left\{ \frac{1}{n} \right\} = 1, \frac{1}{2}, \frac{1}{3}, \ldots$$

and similarly

$$\left\{ \frac{1}{n(n-1)} \right\} = \frac{1}{2 \cdot 1}, \frac{1}{3 \cdot 2}, \frac{1}{4 \cdot 3}, \ldots = \frac{1}{2}, \frac{1}{6}, \frac{1}{12}, \ldots$$

Each of the numbers a_n is said to be a *term* of the sequence $\{a_n\}$. The relevance of the notion of sequence to infinite series is made clear by the following observation. With each infinite series

$$c_0 + c_1 + c_2 + c_3 + \ldots \tag{5}$$

it is possible to associate, in a natural manner, a sequence $\{a_n\}$, where

$$a_0 = c_0, \quad a_1 = c_0 + c_1,$$

and in general,

$$a_n = c_0 + c_1 + c_2 + \cdots + c_n \quad n = 0, 1, 2, 3, \ldots$$

These a_ns are called the *partial sums* of the series (5). It is said that *almost all the terms* of the sequence $\{a_n\}$ have a certain property provided that there is an index n' such that a_n possesses this property whenever $n \geq n'$. For example, almost all the terms of the sequence $\{1/\sqrt{n}\}$ are smaller than 0.1

(here $n' = 101$). Similarly, almost all the terms of the sequence $\{2^n\}$ are greater that 100 (here $n' = 7$).

Let $\{a_n\}$ be a sequence and A a real number. The sequence $\{a_n\}$ is said to *converge* to A if every neighborhood of A contains almost all of the terms of $\{a_n\}$ (Fig. 7.1). For example, the sequence $\{1/n\}$ converges to 0. To see this, let (a, b) be any neighborhood of 0, so that a is necessarily negative and b is positive. To show that (a, b) contains almost all the terms of the given sequence, let n' be any integer greater than $1/b$ (see Proposition 6.3.2). Then, for every integer $n \geq n'$,

$$a < 0 < \frac{1}{n} \leq \frac{1}{n'} < b,$$

so that

$$\frac{1}{n} \in (a, b).$$

Thus, (a, b) contains almost all the terms of the sequence $\{1/n\}$. This particular limit will be used repeatedly in the text and so it is stated as a lemma.

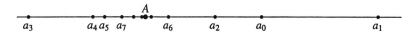

$$\begin{array}{ccccccccc} & & & & A & & & & \\ a_3 & & a_4\, a_5 \;\; a_7 & & a_6 & a_2 & a_0 & & a_1 \end{array}$$

Fig. 7.1   A sequence converging to the limit A.

Lemma 7.1.1 *The sequence* $\{1/n\}$ *converges to* 0.

If the sequence $\{a_n\}$ converges to A, we also say that A is the *limit* of $\{a_n\}$ and write

$$\lim_{n \to \infty} a_n = A \quad \text{or} \quad \{a_n\} \to A \quad \text{or rarely} \quad a_n \to A.$$

Thus, Lemma 7.1.1 can be restated in any of the forms

$$\lim_{n \to \infty} \frac{1}{n} = 0 \quad \text{or} \quad \left\{\frac{1}{n}\right\} \to 0 \quad \text{or} \quad \frac{1}{n} \to 0.$$

It is implicit in the above that every sequence has at most one limit. The formal statement and proof of this intuitively obvious aspect of convergence are relegated to Exercise 4.

A *convergent* sequence is one which has a limit. Thus, $\{1/n\}$ is a convergent sequence. On the other hand, the sequence $\{n\}$ of natural numbers is not convergent since no bounded open interval whatsoever can contain an infinite number of distinct integers. The aforementioned sequence $\{(-1)^n\}$ also fails to converge since any open interval of length 2 or less fails to contain either all of its positive terms or all of its negative terms. A sequence that is not convergent is said to be *divergent*.

It turns out that in order to verify convergence it suffices to check only on neighborhoods centered at the proposed limit.

Lemma 7.1.2 *The sequence $\{a_n\}$ converges to the limit A if and only if every neighborhood of A that is centered at A contains almost all the terms of the sequence.*

Proof. It is clear from Figure 7.2 that if (a, b) is any neighborhood of A and if ε is any positive number that is less than both the positive numbers $A - a$ and $b - A$, then
$$(A - \varepsilon, A + \varepsilon) \subset (a, b).$$
Thus, every neighborhood of A contains a neighborhood of A that is centered at A. Clearly then, if every neighborhood that is centered at A contains almost all the terms of $\{a_n\}$, then so does every neighborhood of A whatsoever. The lemma now follows immediately. $\qquad\square$

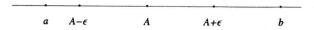

$$a \qquad A{-}\epsilon \qquad\qquad A \qquad\qquad A{+}\epsilon \qquad\qquad b$$

Fig. 7.2 A proof of inclusion.

Given two sequences $\{a_n\}$ and $\{b_k\}$ we say that $\{b_k\}$ is a *subsequence* of $\{a_n\}$ provided for each index k there is an index n_k such that
$$n_0 < n_1 < n_2 < \cdots < n_k < \cdots$$
and
$$b_k = a_{n_k} \qquad \text{for all } k = 0, 1, 2, \ldots.$$
For example, the sequence $\{k^2\} = 0, 1, 4, 9, \ldots$ is a subsequence of $\{n\} = 0, 1, 2, 3, \ldots$. The following proposition should come as no surprise to the reader.

Proposition 7.1.3 *Every subsequence of a convergent sequence $\{a_n\} \to A$ also converges to A.*

Proof. Suppose $\{a_n\} \to A$. If (a, b) is any neighborhood of A, then it contains almost all the terms of $\{a_n\}$. Hence, if $\{a_{n_k}\}$ is any subsequence of $\{a_n\}$, (a, b) contains almost all of its terms too. Consequently $\{a_{n_k}\} \to A$. $\qquad\square$

It is useful to express the geometric condition that every neighborhood of the limit contains almost all the terms of a sequence in a more algebraic manner. This new formulation is now provided.

Theorem 7.1.4 *The sequence $\{a_n\}$ converges to the limit A if and only if for every positive real number ε, there is an index n' such that*
$$|a_n - A| < \varepsilon \qquad \text{for all } n \geq n'. \tag{6}$$

Proof. Let $\varepsilon > 0$. By Lemma 7.1.2, it suffices to prove that $|a_n - A| < \varepsilon$ if and only if $a_n \in (A - \varepsilon, A + \varepsilon)$. This, however, follows from the equivalence of the following statements:

$$|a_n - A| < \varepsilon$$
$$-\varepsilon < a_n - A < \varepsilon$$
$$A - \varepsilon < a_n < A + \varepsilon$$
$$a_n \in (A - \varepsilon, A + \varepsilon) \qquad \square$$

Figure 7.3 illustrates the main argument of the above proposition.

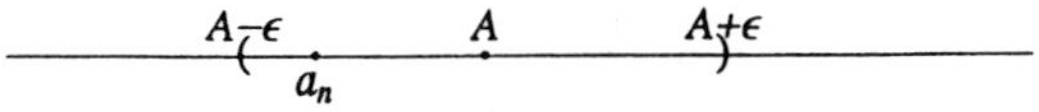

Fig. 7.3   A reformulation of convergence.

Mathematicians often rephrase Equation (6) by saying informally that the *difference $a_n - A$ can be made arbitrarily small*. This phrasing will be used frequently in the sequel and nothing more is to be understood by it than what actually appears in Equation (6). In particular, the convergence $\{a_n\} \to 0$ can be paraphrased by saying that a_n can be made arbitrarily small.

Example 7.1.5 Prove that

$$\lim_{n \to \infty} \frac{2n + 1}{n + 4} = 2.$$

The solutions of problems of this type come frequently in two parts. First, it is useful to examine the desired conclusion and to perform some (occasionally informal) algebraic manipulations that will aid us in producing the final proof. This part is called the *analysis* (no, this has nothing to do with the title of this text). This informal analysis is followed by a formally correct (and preferably succinct) proof whose steps are motivated by the preceding analysis.

Analysis: According to Proposition 7.1.4 it suffices to demonstrate that for every $\varepsilon > 0$, there exists an integer n' such that

$$\left| \frac{2n + 1}{n + 4} - 2 \right| < \varepsilon \qquad \text{for all } n \geq n'. \tag{7}$$

However, Equation (7) is easily seen to be equivalent to

$$\left| \frac{-7}{n + 4} \right| < \varepsilon$$

which, when solved for n, becomes

$$\frac{7}{\varepsilon} - 4 < n.$$

Since all these algebraic manipulations are reversible, it follows that the desired inequality (7) can be guaranteed by restricting attention to those positive integers n that exceed $(7/\varepsilon)-4$, which is clearly true for almost all the positive integers.

Proof. Let ε be any positive real number and let n' be any positive integer greater than $(7/\varepsilon) - 4$. Then, for all $n \geq n'$,

$$\frac{7}{\varepsilon} - 4 < n$$

$$\therefore \quad \frac{7}{\varepsilon} < n + 4$$

$$\therefore \quad \left| \frac{-7}{n+4} \right| < \varepsilon$$

$$\therefore \quad \left| \frac{2n+1}{n+4} - 2 \right| < \varepsilon$$

$$\therefore \quad \lim_{n \to \infty} \frac{2n+1}{n+4} = 2.$$

The above example calls for some comments. The proposed format is only recommended. It could of course be replaced by any other valid proof. Moreover, if this format is followed, it should not be followed blindly. As will be made clear by subsequent examples, the steps of the analysis are in general not reversible and cannot be adhered to too closely in generating the formal proof. They should be viewed merely as indicators of the general outline of the desired proof.

Example 7.1.6 Prove that

$$\lim_{n \to \infty} \frac{2n^2 + 1}{3n^2 - 20} = 2/3.$$

Analysis: To get

$$\left| \frac{2n^2 + 1}{3n^2 - 20} - \frac{2}{3} \right| < \varepsilon$$

solve for n:

$$\left| \frac{43}{3(3n^2 - 20)} \right| < \varepsilon \qquad (8)$$

$$\frac{43}{3\varepsilon} < 3n^2 - 20 \qquad (9)$$

$$\sqrt{\frac{1}{3} \left(\frac{43}{3\varepsilon} + 20 \right)} < n. \qquad (10)$$

Note that the equivalence of (8) and (9) is only guaranteed when $n > 2$; since interest is in large positive values of n, this does not complicate the following proof.

Proof. Let $\varepsilon > 0$ and let n' be any integer greater than both 2 and

$$\sqrt{\frac{1}{3}\left(\frac{43}{3\varepsilon} + 20\right)}.$$

Then, for all $n \geq n'$,

$$\sqrt{\frac{1}{3}\left(\frac{43}{3\varepsilon} + 20\right)} < n$$

$$\therefore \quad \frac{43}{3\varepsilon} < 3n^2 - 20$$

$$\therefore \quad \left|\frac{43}{3(3n^2 - 20)}\right| < \varepsilon$$

$$\therefore \quad \left|\frac{2n^2 + 1}{3n^2 - 20} - \frac{2}{3}\right| < \varepsilon$$

$$\therefore \quad \lim_{n\to\infty} \frac{2n^2 + 1}{3n^2 - 20} = \frac{2}{3}.$$

Exercises 7.1

1. Use the $\varepsilon - n'$ formulation of Theorem 7.1.4 to demonstrate the following convergences:

a) $\left\{\dfrac{1}{n^2}\right\} \to 0$

b) $\left\{\dfrac{1}{2n + 3}\right\} \to 0$

c) $\left\{\dfrac{3n - 2}{4n + 1}\right\} \to \dfrac{3}{4}$

d) $\left\{\dfrac{7n - 5}{5n + 7}\right\} \to \dfrac{7}{5}$

e) $\left\{\dfrac{2n^2}{n^2 + 1}\right\} \to 2$

f) $\left\{\dfrac{2n^2 - 1}{3n^2 + 2}\right\} \to \dfrac{2}{3}$

g) $\left\{\dfrac{n}{n^2 + 1}\right\} \to 0$

h) $\left\{\dfrac{n}{n^2 - 1}\right\} \to 0$

i) $\left\{\dfrac{n}{n^2 + n + 1}\right\} \to 0$

j) $\left\{\dfrac{n}{n^2 - n + 1}\right\} \to 0$

k) $\left\{1 + \dfrac{(-1)^n}{n}\right\} \to 1$

l) $\left\{\dfrac{\sqrt{n} + 1}{3\sqrt{n} - 30}\right\} \to \dfrac{1}{3}$

2. Suppose $\{a_n\} \to A$ and $b_n = (a_1 + a_2 + \cdots + a_n)/n$ for $n = 1, 2, 3, \ldots$. Prove that $\{b_n\} \to A$.

3. Find a non-convergent sequence $\{a_n\}$ such that the sequence

$$\left\{\frac{a_1 + a_2 + \cdots + a_n}{n}\right\}$$

converges.

4. Prove that if $\{a_n\} \to A$ and $\{a_n\} \to B$, then $A = B$.

5. Suppose $\{a_n\}$ is a sequence such that $|a_n - a_m| \leq 1/|m - n|$ for all indices m, n. Prove that $a_0 = a_1 = a_2 = \cdots = a_n = \cdots$.

6. Let A be an upper bound of a set S of real numbers. Prove that A is the least upper bound of S if and only if one of the following two conditions hold:
   a) $A \in S$
   b) A is the limit of some sequence $\{a_n\} \subset S$.

7. Suppose $\alpha = \{a_n\}$ and $\beta = \{b_n\}$ are two sequences. Construct at least five different sequences whose terms include those of both α and β.

8. Suppose $\alpha = \{a_n\}$, $\beta = \{b_n\}$, and $\gamma = \{c_n\}$ are three sequences. Construct a sequence whose terms include all the terms of α, β, and γ.

9. Suppose k is a positive integer and

$$\alpha_1 = \{a_{1,n}\}_{n=0}^{\infty}, \alpha_2 = \{a_{2,n}\}_{n=0}^{\infty}, \ldots, \alpha_k = \{a_{k,n}\}_{n=0}^{\infty}$$

are k sequences. Prove that there is a sequence whose terms include all those of $\alpha_1, \alpha_2, \ldots \alpha_k$, i.e., prove that there is a sequence whose terms include all the terms of the form $a_{i,n}$ for $i = 1, 2, \ldots, k$ and $n = 0, 1, 2, \ldots$.

10. Suppose $a_k = \{a_{k,n}\}_{n=0}^{\infty}$ is a sequence for each positive integer $k = 0, 1, 2, \ldots$. Prove that there is a sequence whose terms include all the terms of all these sequences.

A set is said to be denumerable if there is a sequence that contains all the elements of the set. Prove that the following sets are all denumerable.

11. The set of all the positive rational numbers.

12. The set of all the rational numbers.

13. The set of all solutions of equations of the form $ax = b$ where a and b are integers.

14. The set of all solutions of equations of the form $ax^2 + bx + c$ where a, b, c are integers.

15. The set of all solutions of equations of the form $a_0x^n + a_1x^{n-1} + \ldots + a_n = 0$ where n is a positive integer and $a_0, a_1, a_2, \ldots, a_n$ are arbitrary integers.

16. Prove that the set of real numbers is not denumerable.

17. Describe the main mathematical achievements of Augustin–Louis Cauchy.

7.2   LIMIT THEOREMS

We are about to present a theorem whose several parts are very useful in the computation of the limits of more complicated sequences. The proof of this theorem requires two lemmas.

Lemma 7.2.1 *If $\{a_n\} \to A \neq 0$ then*

$$|a_n| > \frac{|A|}{2} \qquad \text{for almost all } n.$$

Proof. Suppose first that A is negative. By Theorem 7.1.4 the open interval $(3A/2, A/2)$ contains almost all the terms of the given sequence. Hence there exists an integer n' such that

$$a_n \in \left(\frac{3A}{2}, \frac{A}{2}\right) \qquad \text{for all } n \geq n'.$$

However, it is clear from Figure 7.4 that any a_n in $(3A/2, A/2)$ must satisfy $|a_n| > |A|/2$ and so we are done.

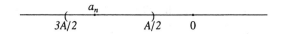

Fig. 7.4 Bounding a sequence away from 0.

The argument in the case that A is positive is similar (Exercise 26). □

Lemma 7.2.2 *Every convergent sequence is bounded.*

Proof. Suppose $\{a_n\} \to A$. It follows from the definition of convergence that the open interval $(A-1, A+1)$ contains all the terms $\{a_n\}_{n=n'}^{\infty}$ for some index n'. Since

$$\{a_n\}_{n=0}^{\infty} = \{a_0, a_1, a_2, \ldots, a_{n'-1}\} \cup \{a_n\}_{n=n'}^{\infty}$$

it follows from Lemma 6.2.4 that the given sequence is bounded. □

Theorem 7.2.3 *Suppose $\{a_n\} \to A$, $\{b_n\} \to B$ and r is a real number. Then*

1. $\lim_{n\to\infty} r = r$

2. $\lim_{n\to\infty}(a_n + b_n) = A + B$

3. $\lim_{n\to\infty}(a_n b_n) = A \cdot B$

4. $\lim_{n\to\infty}(a_n - b_n) = A - B$

5. $\lim_{n\to\infty} a_n/b_n = A/B$ *provided $B \neq 0$.*

Proof of 1. As every neighborhood of r contains <u>all</u> the terms of the sequence $\{r\}$ it follows immediately that $\{r\} \to r$.

Analysis of 2: We need

$$|(a_n + b_n) - (A + B)| < \varepsilon. \tag{11}$$

Since

$$|(a_n + b_n) - (A + B)| = |(a_n - A) + (b_n - B)| \le |a_n - A| + |b_n - B|$$

can be insured provided that

$$|a_n - A| < \frac{\varepsilon}{2} \quad \text{and} \quad |b_n - B| < \frac{\varepsilon}{2}$$

separately, which is easy enough to guarantee since $\{a_n\} \to A$ and $\{b_n\} \to B$.

Proof of 2. Let $\varepsilon > 0$. Because $\{a_n\} \to A$ and $\{b_n\} \to B$,

$$\text{there exists } n' \text{ such that } |a_n - A| < \frac{\varepsilon}{2} \text{ for all } n \ge n'$$

and

$$\text{there exists } n'' \text{ such that } |b_n - B| < \frac{\varepsilon}{2} \text{ for all } n \ge n''.$$

Let $n^* = \max\{n', n''\}$. Then for all $n \ge n^*$,

$$\begin{aligned}
|(a_n + b_n) - (A + B)| &= |(a_n - A) + (b_n - B)| \\
&\le |a_n - A| + |b_n - B| < \frac{\varepsilon}{2} + \frac{\varepsilon}{2} < \varepsilon.
\end{aligned}$$

Hence, by Theorem 7.1.4, $\{a_n + b_n\} \to A + B$.

Analysis of 3. We need to insure that

$$|a_n b_n - AB| < \varepsilon.$$

By adding and subtracting the same term (one of the most common and underhanded of all mathematical tricks), we get

$$\begin{aligned}
|a_n b_n - AB| &= |(a_n - A)b_n + A(b_n - B)| \\
&\le |(a_n - A)b_n| + |A(b_n - B)| \\
&= |a_n - A| \cdot |b_n| + |A| \cdot |b_n - B| \tag{12}
\end{aligned}$$

By Lemma 7.2.2, $\{b_n\}$ is bounded, so that there exists a number M such that $|b_n| < M$ for all n. This fact, together with the convergences $\{a_n\} \to A$ and $\{b_n\} \to B$ is sufficient to guarantee that (12) can be made smaller than ε.

Proof of 3. Let $\varepsilon > 0$ and let M be as above. Because $\{a_n\} \to A$ and $\{b_n\} \to B$,

$$\text{there exists } n' \text{ such that } |a_n - A| < \frac{\varepsilon}{2M} \text{ for all } n \ge n'$$

and

$$\text{there exists } n'' \text{ such that } |b_n - B| < \frac{\varepsilon}{2(|A| + 1)} \text{ for all } n \ge n''.$$

The reason for using $|A| + 1$ instead of $|A|$ in the definition of n'' is that A could conceivably be zero in which case it could not be used as a factor in the denominator. The quantity $|A| + 1$, on the other hand, is of course not zero.

Let $n^* = \max\{n', n''\}$. Then for all $n \geq n^*$,

$$|a_n b_n - AB| = |(a_n - A)b_n + A(b_n - B)|$$
$$\leq |a_n - A| \cdot |b_n| + |A| \cdot |b_n - B| < \frac{\varepsilon}{2M} \cdot M + |A| \cdot \frac{\varepsilon}{2(|A| + 1)}$$
$$< \frac{\varepsilon}{2} + \frac{\varepsilon}{2} = \varepsilon.$$

Hence, by Theorem 7.1.4, $\{a_n b_n\} \to AB$.

Proof of 4. By 1–3,

$$\lim_{n \to \infty} (a_n - b_n) = \lim_{n \to \infty} [a_n + (-1)b_n] = \lim_{n \to \infty} a_n + \lim_{n \to \infty} [(-1)b_n]$$
$$= \lim_{n \to \infty} a_n + \lim_{n \to \infty} (-1) \lim_{n \to \infty} b_n = \lim_{n \to \infty} a_n - \lim_{n \to \infty} b_n.$$

Analysis of 5. In view of 3 and the fact that

$$\frac{a_n}{b_n} = a_n \cdot \frac{1}{b_n},$$

it suffices to prove that

$$\lim_{n \to \infty} \frac{1}{b_n} = \frac{1}{B}.$$

To accomplish this we need

$$\left| \frac{1}{b_n} - \frac{1}{B} \right| < \varepsilon$$

or

$$\left| \frac{B - b_n}{b_n B} \right| < \varepsilon.$$

Since $\{b_n\} \to B$, the numerator can of course be made small. At the same time, however, it is necessary to insure that the denominator does not become too small. The reason for the latter concern is that too small a denominator would again increase the numerical value of the fraction. Lemma 7.2.1 guarantees that the denominator stays at least as large as $|B|^2/2$.

Proof of 5. Let $\varepsilon > 0$. Because $\{b_n\} \to B \neq 0$,

there exists n' such that $|B - b_n| < \dfrac{\varepsilon B^2}{2}$ for all $n \geq n'$

and

there exists n'' such that $|b_n| > \dfrac{B}{2}$ for all $n \geq n''$.

Set $n^* = \max\{n', n''\}$. Then, for all $n \geq n^*$

$$\left| \frac{1}{b_n} - \frac{1}{B} \right| = \frac{|B - b_n|}{|B| \cdot |b_n|} < \frac{\dfrac{\varepsilon B^2}{2}}{\dfrac{|B|}{2} \cdot |B|} = \varepsilon$$

and hence $\{1/b_n\} \to 1/B$. It now follows from part 3 that

$$\lim_{n \to \infty} \frac{a_n}{b_n} = \left(\lim_{n \to \infty} a_n \right) \left(\lim_{n \to \infty} \frac{1}{b_n} \right) = A \cdot \frac{1}{B} = \frac{A}{B}. \qquad \square$$

Example 7.2.4 Compute

$$\lim_{n \to \infty} \frac{3n^3 - 5n^2 + 1}{2n^3 + 4n - 17}.$$

Divide both the numerator and denominator of the general term by n^3 so as to transform it to the form

$$\frac{3 - 5 \cdot \left(\dfrac{1}{n} \right)^2 + \left(\dfrac{1}{n} \right)^3}{2 + 4 \cdot \dfrac{1}{n} - 17 \cdot \left(\dfrac{1}{n} \right)^3},$$

which, in view of Lemma 7.1.1 and the several parts of Theorem 7.2.3 converges to

$$\frac{3 - 5 \cdot 0^2 + 0^3}{2 + 4 \cdot 0 - 17 \cdot 0^3} = \frac{3}{2}.$$

It is customary to think of the next proposition as a *squeezing principle*. It greatly simplifies the computation of the limits of sequences with complicated terms.

Proposition 7.2.5 *Suppose* $\lim_{n \to \infty} a_n = \lim_{n \to \infty} b_n = A$ *and*

$$a_n \leq c_n \leq b_n \qquad \text{for all } n = 0, 1, 2, 3, \dots .$$

Then $\lim_{n \to \infty} c_n = A$.

Proof. Let (a, b) be any neighborhood of A. Then (a, b) contains almost all the a_n and almost all the b_n. It follows from the assumption $a_n \leq c_n \leq b_n$ that (a, b) also contains c_n whenever it contains both a_n and b_n. Thus, (a, b) contains almost all the c_n and hence $\{c_n\} \to A$. $\qquad \square$

Example 7.2.6 Prove that

$$\lim_{n \to \infty} \frac{\sin(n^2 - n)}{n} = 0.$$

For any x, $|\sin x| \leq 1$ and hence

$$-\frac{1}{n} \leq \frac{\sin(n^2 - n)}{n} \leq \frac{1}{n}.$$

Since

$$\lim_{n \to \infty} \left(-\frac{1}{n}\right) = \lim_{n \to \infty} \frac{1}{n} = 0,$$

it follows from Proposition 7.2.5 that

$$\lim_{n \to \infty} \frac{\sin(n^2 - n)}{n} = 0.$$

Proposition 7.2.7 *Let a be a fixed real number. Then*

$$\lim_{n \to \infty} a^n \begin{cases} = 0 & \text{if } |a| < 1 \\ = 1 & \text{if } a = 1 \\ \text{does not exist} & \text{if } |a| > 1 \\ \text{does not exist} & \text{if } a = -1 \end{cases}$$

Proof. Suppose first that $0 < a < 1$. It follows that $1/a > 1$ and so we may write

$$\frac{1}{a} = 1 + h \qquad \text{where } h > 0.$$

Hence, by Proposition 6.2.2.12,

$$0 < a^n = \frac{1}{(1+h)^n} \leq \frac{1}{1 + nh} < \frac{1}{h} \cdot \frac{1}{n}.$$

Since the last term converges to $(1/h) \cdot 0 = 0$, it follows from Proposition 7.2.5 that $\lim_{n \to \infty} a^n = 0$.

If $-1 < a < 0$, then

$$-|a|^n \leq a^n \leq |a|^n.$$

However, by the first part of the proof,

$$\lim_{n \to \infty} |a|^n = 0 = \lim_{n \to \infty} (-|a|^n)$$

and so by Proposition 7.2.5 we conclude that $\lim_{n \to \infty} a^n = 0$ again.

It is obvious that $\{0^n\} \to 0$ and so we are done with the case $|a| < 1$.

That $\{1^n\} \to 1$ is also obvious. Since no neighborhood of length 2 or less contains both 1 and -1, it follows that the sequence $\{(-1)^n\}$ does not converge. Finally, if $|a| > 1$, then, by Proposition 6.2.2.12 we have

$$a^{2n} = [1 + (a^2 - 1)]^n \geq 1 + n(a^2 - 1)$$

and so $\{a^n\}$ is unbounded and does not converge. $\qquad \square$

The following propositions are quite plausible and their formal proofs are relegated to Exercises 15 to 23. They will be taken for granted in the sequel.

Proposition 7.2.8 $\{a_n\} \to A$ *if and only if* $\{a_n - A\} \to 0$.

Proposition 7.2.9 *If* $\{a_n\} \to A$ *then* $\{|a_n|\} \to |A|$.

Proposition 7.2.10 *Suppose* $\{a_n\} \to A$ *and* $\{b_n\} \to B$.

1. *If* $a_n \geq 0$ *for all* n *then* $A \geq 0$.

2. *If* $a_n \geq b_n$ *for all* n *then* $A \geq B$.

Proposition 7.2.11 *Suppose* $\{a_n\}$ *is a sequence and* $\{a_{n_m}\}$, $\{a_{n_k}\}$ *are two subsequences whose union contains all the terms of* $\{a_n\}$. *Then* $\{a_n\} \to A$ *if and only if both* $\{a_{n_m}\} \to A$ *and* $\{a_{n_k}\} \to A$.

Proposition 7.2.12 *If* $a_n \geq 0$ *for each* n, r *is a rational number,* $\{a_n\} \to A$, *and* A^r *exists, then* $\{a_n^r\} \to A^r$.

Proposition 7.2.13 *The convergence of a sequence is not affected by the deletion, addition, or modification of a finite number of terms.*

Proposition 7.2.14 *Let* s *be a fixed integer. Then*

$$\lim_{n \to \infty} n^s \begin{cases} = 0 & \text{if } s < 0 \\ = 1 & \text{if } s = 0 \\ \text{does not exist} & \text{if } s > 0. \end{cases}$$

Proposition 7.2.15 *Suppose* a_0 *and* b_0 *are non-zero real numbers, and* d *and* e *are non-negative integers. Then*

$$\lim_{n \to \infty} \frac{a_0 n^d + a_1 n^{d-1} + a_2 n^{d-2} + \cdots + a_d}{b_0 n^e + b_1 n^{e-1} + b_2 n^{e-2} + \cdots + b_e} = \begin{cases} 0 & \text{if } d < e \\ \dfrac{a_0}{b_0} & \text{if } d = e \end{cases}$$

and this limit does not exist if $d > e$.

Proposition 7.2.16 *Let* r *be a real number. Then*

1. *There is a sequence of distinct rational numbers* $\{a_n\}$ *that converges to* r.

2. *There is a sequence of distinct irrational numbers* $\{b_n\}$ *that converges to* r.

Example 7.2.17 Show that $\lim_{n\to\infty}(\sqrt{n+2}-\sqrt{n}) = 0$. Observe that

$$0 \leq \sqrt{n+2} - \sqrt{n} = \frac{(\sqrt{n+2} - \sqrt{n}) \cdot (\sqrt{n+2} + \sqrt{n})}{\sqrt{n+2} + \sqrt{n}}$$

$$= \frac{(n+2) - n}{\sqrt{n+2} + \sqrt{n}} < \frac{2}{\sqrt{n}}.$$

However, it follows from Propositions 7.1.1 and 7.2.12 that $1/\sqrt{n} \to 0$ and hence, by Proposition 7.2.5, $\lim_{n\to\infty}(\sqrt{n+2}-\sqrt{n}) = 0$.

Exercises 7.2

1. Evaluate the limits of the following sequences when they converge, or else demonstrate their divergence.

a) $\left\{ \dfrac{3n^2 + 5n + 1}{n + 1} \right\}$ b) $\left\{ \dfrac{3n^2 + 5n + 1}{n^2 + 1} \right\}$

c) $\left\{ \dfrac{3n^2 + 5n + 1}{n^3 + 1} \right\}$ d) $\dfrac{.5^n + 3\sin n}{\sqrt{n}}$

e) $\left\{ \dfrac{2^n - 1}{3^n + 1} \right\}$ f) $\left\{ \dfrac{2^n + 1}{3^n - 1} \right\}$

g) $\left\{ \dfrac{3^n + 2^n}{3^n - 2^n} \right\}$ h) $\left\{ \dfrac{3^n + 4^{n-3}}{5^{n+2} - 2^{n+4}} \right\}$

i) $\left\{ \dfrac{4^{2n-3} + 2^{5n+6}}{5^{3n-2} - 3^{n+10}} \right\}$ j) $\{n^n\}$

k) $\left\{ \dfrac{1}{n^n} \right\}$ l) $\left\{ 1 - \left| \dfrac{\sin n}{n} \right| \right\}$

m) $\left\{ \dfrac{n}{2^n} \right\}$ n) $\left\{ \dfrac{n^2}{2^n} \right\}$

o) $\left\{ \dfrac{n!}{n^n} \right\}$ p) $\left\{ \dfrac{n!}{2^n} \right\}$

2. Suppose $\{a_n\} \to A$ and $a_n \geq 5$ for each $n \geq 127$. Prove that $A \geq 5$.

3. Suppose $\{a_n\} \to A$ and $a_n \geq 0$ for all n. Prove that $\{\sqrt{n+a_n} - \sqrt{n}\} \to 0$.

4. Prove that if $a \geq 0$, then $\lim_{n\to\infty}(\sqrt{n+2a} - \sqrt{n}) \cdot \sqrt{n+a} = a$.

5. Suppose $\{a_n\} \to 0$ and $a_n \neq 0$ for all n. Prove that the sequence $\{1/a_n\}$ diverges.

6. Does the sequence $\{1/(\sqrt[3]{n+1} - \sqrt[3]{n})\}$ converge or diverge?

7. Show that if $\{a_n\} \to A$ and $b_n = (a_n + a_{n+1})/2$ for all n, then $\{b_n\} \to A$.

8. Suppose $\{a_n\} \to A$ and $b_n = (a_n + a_{n+1})/3$ for all n. Does $\{b_n\}$ converge or diverge? Prove your answer.

9. Suppose $\{a_n\} \to A$ and $b_n = (a_n + a_{n+1})/n$ for all n. Does $\{b_n\}$ converge or diverge? Prove your answer.

10. Suppose $\{a_n\} \to 1$ and $b_n = na_n + n^2 a_{n+1}$ for all n. Does $\{b_n\}$ converge or diverge? Prove your answer.

11. Suppose $\{a_n\} \to 0$ and $b_n = na_n + n^2 a_{n+1}$ for all n. Does $\{b_n\}$ converge or diverge? Prove your answer.

12. Suppose $\{a_n\} \to A$. Evaluate

$$\lim_{n \to \infty} \frac{1}{n^p} \sum_{k=0}^{n} a_n$$

for

    a) $\ p = 1$ b) $\ p < 1$ c) $\ p > 1$.

13. Find a divergent sequence $\{a_n\}$ such that

$$\frac{1}{n^p} \sum_{k=0}^{n} a_k$$

converges, where $p > 0$.

14. Let a be a fixed non-negative real number. Prove that $\lim_{n \to \infty} \sqrt[n]{a} = 1$. (Hint: Assume first that $a > 1$, and use Proposition 1.1.2 to argue that

$$0 < a^{1/n} - 1 = \frac{a - 1}{a^{(n-1)/n} + a^{(n-2)/n} + \cdots + a^{1/n} + 1} < \frac{a-1}{n}.)$$

15. Prove Proposition 7.2.8.

16. Prove Proposition 7.2.9. (Hint: use Exercise 6.2.34.c)

17. Prove Proposition 7.2.10.

18. Prove Proposition 7.2.11.

19. Prove Proposition 7.2.12.

20. Prove Proposition 7.2.13.

21. Prove Proposition 7.2.14.

22. Prove Proposition 7.2.15.

23. Prove Proposition 7.2.16.

24. Suppose the variable s of Proposition 7.2.14 is allowed to range over all rational numbers. Prove that the proposition remains valid.

25. Prove that $\lim_{n\to\infty} \sqrt[n]{n} = 1$. (Hint: Set $s_n = \sqrt[n]{n} - 1$ and use the Binomial Theorem to prove that $n > \binom{n}{2} s_n^2$ for $n \geq 2$.)

26. Complete the proof of Proposition 7.2.1.

27. Use the Binomial Theorem to prove that if $0 < a < 1$, then $\lim_{n\to\infty} na^n = 0$. (Hint: Write $a = 1/(1+h)$ for some $h > 0$.)

28. Suppose $0 < a < 1$. Prove that for every integer k, $\lim_{n\to\infty} n^k a^n = 0$.

29. Find two sequences $\{a_n\} \to A$ and $\{b_n\} \to B$ such that $a_n < b_n$ for all n, but $A \not< B$.

30. Suppose $a \geq b > 0$, and $x_n = [(a^n + b^n)/2]^{1/n}$ for $n = 1, 2, 3, \ldots$. Prove that $\{x_n\} \to a$.

31. Suppose $\{a_n\}$ is a sequence of rational numbers that converges to 0. Prove that for every real number r, $\{r^{a_n}\} \to 1$.

32. Suppose $a_n \in [a, b]$ for $n = 0, 1, 2, \ldots$ and $\{a_n\} \to A$. Prove that $A \in [a, b]$.

33. Prove that

$$\lim_{m\to\infty} \left(\lim_{n\to\infty} \frac{mx}{m+n} \right) = 0$$

whereas

$$\lim_{n\to\infty} \left(\lim_{m\to\infty} \frac{mx}{m+n} \right) = x.$$

34. Prove that

$$\lim_{n\to\infty} \left(\lim_{m\to\infty} (\cos^2(n!\pi x))^m \right) = 1 \text{ or } 0$$

according to whether x is rational or not.

35. Suppose $a_{n+1} = (a_n - 1)/2$ for $n = 0, 1, 2, \ldots$. Prove that $\{a_n\} \to -1$ (regardless of the value of a_0).

36. Suppose r is a real number and $a_{n+1} = (a_n + a)/2$ for $n = 0, 1, 2, \ldots$. Prove that $\{a_n\}$ is a convergent sequence.

Chapter Summary

Sequences and their limits are defined. Various propositions are stated and proved for the purpose of facilitating the evaluation of the limits of many sequences.

8

_____

The Cauchy Property

So far, proving that a series is convergent is synonymous with evaluating its limit. This is too restrictive a state of affairs. This chapter provides two criteria that facilitate proofs of convergence of sequences even when their limits remain unknown.

8.1 LIMITS OF MONOTONE SEQUENCES

While we aim to eventually discuss all sequences in this chapter, it is convenient to first focus on a restricted class of sequences. A sequence $\{a_n\}$ is said to be _increasing_ if $a_n \leq a_{n+1}$ for every n. It is said to be _decreasing_ if $a_n \geq a_{n+1}$ for every n. A sequence is said to be _monotone_ if it is either increasing or decreasing. Thus, the sequence $\{2^n\}$ is increasing, the sequence $\{1/n\}$ is decreasing, and the sequence $\{(-1)^n\}$ is not monotone. The following theorem, which merely asserts the existence of the limits of certain monotone sequences, turns out to be very useful both in evaluating specific limits and in facilitating the proofs of fundamental theorems. In order to emphasize the significance of this theorem, its proof is postponed until after two of its important implications have been discussed.

Theorem 8.1.1 (Monotone convergence) _Every bounded monotone sequence converges._

Figure 8.1 below illustrates this theorem with an increasing sequence $\{a_n\}$ that is bounded above by M. According to the theorem this sequence converges to a limit that must, of course, lie between each a_n and M.

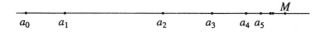

Fig. 8.1 A bounded increasing sequence

As an illustration of the utility of this theorem, we remind the readers that it is customary (and correct) to identify the real numbers with their decimal expansions. It is the Monotone Convergence Theorem that justifies the tacit assumption that every decimal expansion of the form

$$m.d_1d_2d_3\ldots = m + \frac{d_1}{10} + \frac{d_2}{10^2} + \frac{d_3}{10^3} + \cdots \qquad (1)$$

does indeed represent a real number. Note that if a_n denotes the finite decimal expansion

$$m.d_1d_2\ldots d_n = m + \frac{d_1}{10} + \frac{d_2}{10^2} + \cdots + \frac{d_n}{10^n}$$

then $\{a_n\}$ is clearly an increasing sequence which is bounded between m and $m+1$ (Exercise 44). By the Monotone Convergence Theorem this sequence converges, and its limit A is taken to be the value of the infinite decimal expansion of Equation (1).

In addition to this justification of a widely held belief, the Monotone Convergence Theorem also has other, more surprising, applications.

Example 8.1.2 Consider the recursively defined sequence

$$a_0 = 1$$
$$a_{n+1} = \sqrt{1 + a_n} \qquad n = 0, 1, 2, 3 \ldots .$$

Its first few terms are

$$a_0 = 1,$$
$$a_1 = \sqrt{2} = 1.41\ldots ,$$
$$a_2 = \sqrt{1 + \sqrt{2}} = 1.55\ldots ,$$
$$a_3 = \sqrt{1 + \sqrt{1 + \sqrt{2}}} = 1.59\ldots ,$$

indicating that the sequence is increasing. This monotonicity is easily demonstrated by induction on n. It is already known that $a_0 < a_1$ and the assumption $a_n < a_{n+1}$ leads to

$$1 + a_n < 1 + a_{n+1}$$
$$\therefore \quad \sqrt{1 + a_n} < \sqrt{1 + a_{n+1}}$$

$$\therefore \quad a_{n+1} < a_{n+2}.$$

Thus, the inequality $a_n < a_{n+1}$ holds for all n and the given sequence is indeed monotone increasing. Since all of the terms of the sequence are positive, finding an upper bound will demonstrate its boundedness. Once again, an easy inductive argument demonstrates that the sequence is bounded above by 2. For clearly $a_0 < 2$, and the hypothesis $a_n < 2$ leads to the conclusion

$$a_{n+1} = \sqrt{1 + a_n} < \sqrt{1 + 2} < 2.$$

Thus, $a_n < 2$ for all n. We conclude from the Monotone Convergence Theorem that the given sequence is in fact convergent. Its limit can of course be estimated to any desired accuracy by simply evaluating further and further terms of the sequence, and in many similar cases that is the best that can be expected. In this case, however, more information can be obtained about the limit. Suppose $\lim_{n \to \infty} a_n = A$. It then follows from the definition of the given sequence that

$$A^2 = \lim_{n \to \infty} a_n^2 = \lim_{n \to \infty} (1 + a_{n-1}) = 1 + \lim_{n \to \infty} a_{n-1} = 1 + A.$$

Hence, $A^2 - A - 1 = 0$, or

$$A = \frac{1 \pm \sqrt{5}}{2}.$$

Since $\{a_n\}$ is an increasing sequence of positive numbers it follows that A must also be positive and hence

$$\lim_{n \to \infty} a_n = A = \frac{1 + \sqrt{5}}{2}.$$

We now return to the Monotone Convergence Theorem. This theorem does not hold for the rational numbers system. For example, the sequence of rational numbers

$$a_0 = 1, \quad a_1 = 1.4, \quad a_2 = 1.41, \quad a_3 = 1.414, \quad a_4 = 1.4142, \quad \ldots$$

whose terms consist of progressively longer initial segments of the decimal expansion of $\sqrt{2} = 1.41421356\ldots$ does not converge to a rational number, since it converges, by its definition, to an irrational number, namely $\sqrt{2}$. It follows that this highly desirable theorem cannot be proven on the basis of the Field Axioms F1 to F7, O1 to O5 (Section 6.2) alone. Another axiom is needed; it is the Completeness Axiom (Chapter 6) which states

every non-empty set of real numbers that is bounded above has a least upper bound.

Proof of Theorem 8.1.1. Let $\{a_n\}$ be a bounded monotone sequence. It will be shown that if $\{a_n\}$ is increasing then it converges, the other case being relegated to Exercise 31.

It follows from the Completeness Axiom that $\{a_n\}$ has a least upper bound, say b, so that we have

$$a_0 \le a_1 \le a_2 \le \ldots \le a_n \le \ldots \le b. \qquad (2)$$

If (c, d) is any neighborhood of b, then (c, d) must contain a term $a_{n'}$ of $S = \{a_n\}$, for otherwise c would be a smaller upper bound of S than b. But then $a_n \in (c, d)$ for all $n \ge n'$. Hence, $\{a_n\} \to b$. □

Example 8.1.3 Suppose $a_0 = 1$ and $a_{n+1} = a_n + (1/a_n)$ for all n. Does the sequence $\{a_n\}$ converge or diverge? It is clear that this sequence is monotone increasing but it is not clear whether or not it is bounded. The first six terms

$$a_0 = 1,$$
$$a_1 = 2,$$
$$a_2 = 2.5,$$
$$a_3 = 2.9,$$
$$a_4 = 3.24\ldots,$$
$$a_5 = 3.55\ldots$$

are not helpful in this regard. The sequence is in fact _not_ convergent. For suppose A were its limit. Since $\{a_n\}$ is increasing, $A > 0$. It now follows from the definition of the sequence and Theorem 7.2.3 that

$$A = \lim_{n \to \infty} a_{n+1} = \lim_{n \to \infty} a_n + \frac{1}{\lim\limits_{n \to \infty} a_n} = A + \frac{1}{A}$$

or

$$0 = \frac{1}{A}$$

which is of course impossible.

The following easy consequence of the Monotone Convergence Theorem turns out to be useful in a surprising number of cases.

Proposition 8.1.4 _Suppose_ $\{a_n\}$ _is a monotone increasing sequence and_ $\{b_n\}$ _is a monotone decreasing sequence such that_

$$a_n \le b_n \qquad for\ all\ n = 0, 1, 2, \ldots$$

and

$$\{a_n - b_n\} \to 0.$$

Then the two sequences converge to the same limit.

Proof. Since $a_0 \le a_n \le b_n \le b_0$ it follows that both sequences are contained in the interval $[a_0, b_0]$ and hence both are bounded. By the Monotone Convergence Theorem, both sequences converge. Now that it is known that both of the sequences converge, the equality of their limits follows from Proposition 7.2.3.4 and the fact that $\{a_n - b_n\} \to 0$. $\qquad\qquad\qquad\square$

Monotone sequences owe their theoretical significance to the next proposition which demonstrates that these sequences, despite their special nature, are in fact ubiquitous.

Proposition 8.1.5 *Every sequence has a monotone subsequence.*

Proof. Let $\{a_n\}$ be a sequence. A term $a_{n'}$ is said to have an *open view* (of infinity) if

$$a_{n'} > a_n \qquad \text{for all } n \geq n'. \qquad (3)$$

In other words, a term has an open view if it exceeds every subsequent term. Every sequence clearly contains either an infinite or a finite number of terms with an open view.

Case 1. The sequence $\{a_n\}$ has an infinite number of terms with an open view, say $a_{n_0}, a_{n_1}, a_{n_2}, \ldots$ where $n_0 < n_1 < n_2 \ldots$. In this case it follows from Inequality (3) that

$$a_{n_0} > a_{n_1} > a_{n_2} > \cdots$$

so that the terms $\{a_{n_k}\}_{k=0}^{\infty}$ form the required monotone subsequence of $\{a_n\}$.

Case 2. The sequence $\{a_n\}$ has only a finite number of terms with an open view. Consequently, there is an index n^* such that for every $n \geq n^*$ the term a_n does not have an open view, that is

$$\text{for each } n \geq n^* \text{ there exists an index } m > n \text{ such that } a_n \leq a_m. \qquad (4)$$

Set $n_0 = n^*$. Statement (4) guarantees that there exists an index $n_1 > n_0$ such that $a_{n_0} \leq a_{n_1}$. As $n_1 > n^*$ there again exists an index $n_2 > n_1$ such that $a_{n_1} \leq a_{n_2}$. Since this process can be repeated indefinitely, it follows that there exist indices $n_0 < n_1 < n_2 < \ldots$ such that

$$a_{n_0} \leq a_{n_1} \leq a_{n_2} \leq \cdots.$$

Consequently, $\{a_{n_k}\}_{k=0}^{\infty}$ is the required monotone subsequence of $\{a_n\}$. $\qquad\square$

Example 8.1.6 The sequence $\{n\}$ contains a monotone increasing sequence but no monotone decreasing sequence. The sequence $\{1/n\}$ contains a monotone decreasing sequence but no monotone increasing sequence. Let d_n denote the digit in the nth place in the decimal expansion of $1/7 = .\overline{142857}$. The sequence $\{d_n\}$ has the constant (and hence monotone) subsequence $1, 1, 1, \ldots$ as a subsequence, as well as five others. If e_n is the digit in the nth place of the decimal expansion of $\sqrt{2}$, then $\{e_n\} = 1, 4, 1, 4, 2, 1, 3, 5, 6, 2, 3, 7, 3, 0, 9, 5, 0, 4,$ $8, 8, \ldots$. Since the decimal expansion of $\sqrt{2}$ is not periodic the previous argument fails to prove the existence of a constant subsequence of $\{e_n\}$. Nevertheless, the existence of such a subsequence can argued as follows. There are only 10 digits $0, 1, 2, \ldots, 9$ that are used in each of the infinitely many decimal places in the expansion of $\sqrt{2}$. Hence some digit e must be occur infinitely many times so that $e, e, e, \ldots$ is a constant subsequence of $\{e_n\}$. Unfortunately, while it seems likely that <u>every</u> one of the digits $0, 1, 2, \ldots, 9$ occurs infinitely many times in the decimal expansion of $\sqrt{2}$, it is beyond the power of mathematicians today to produce even one digit of which this can be asserted with mathematical certainty. Consequently, the exact nature of the monotone subsequences of $\{e_n\}$ is unknown. Finally, we define inductively a new sequence $\{a_n\}$ where a_n denotes the difference between the number of odd and the number of even digits in the set $\{e_0, e_1, e_2, \ldots, e_n\}$. Thus, $\{e_n\} = 1, 0, 1, 0, -1, 0, 1, 2, 1, 0, 1, 2, 3, 2, 3, 4, 3, 2, 1, 0, \ldots$. While the previous proposition asserts that this sequence contains a monotone subsequence, it is unknown whether or not it contains a monotone increasing sequence.

The following corollary will be used repeatedly in the proofs of subsequent theorems. As the existence of its title implies, this is generally recognized as one of the more important properties of the real number system, and so it also called a theorem.

Theorem 8.1.7 (Bolzano–Weierstrass) *Every bounded sequence has a convergent subsequence.*

Proof. By the previous proposition every sequence has a monotone subsequence. Since the boundedness of the original sequence also guarantees the boundedness of the monotone subsequence, it follows from the Completeness Axiom that this monotone subsequence is convergent. ☐

Exercises 8.1

In Exercises 1–16, 18–26 decide whether the given sequence converges or diverges. Prove your answer and compute the limit if it exists.

1. $a_0 = 1$ and $a_{n+1} = \sqrt{1 + a_n^2}$ for $n = 0, 1, 2, \ldots$.

2. $a_0 = 1$ and $a_{n+1} = \sqrt{1 + 2a_n}$ for $n = 0, 1, 2, \ldots$.

3. $a_0 = 1$ and $a_{n+1} = 1 + \sqrt{a_n}$ for $n = 0, 1, 2, \ldots$.

4. $a_0 = 1$ and $a_{n+1} = \sqrt{3 + 2a_n}$ for $n = 0, 1, 2, \ldots$.

5. $a_0 = 1$ and $a_{n+1} = \sqrt{a_n + \dfrac{1}{a_n}}$ $n = 0, 1, 2, \ldots$.
   (Hint: Show first that $a_n > 1$.)

6. $a_0 = 1$ and $a_{n+1} = 2 - \dfrac{1}{a_n}$ for $n = 0, 1, 2, \ldots$.

7. $a_0 = 1$ and $a_{n+1} = 3 - \dfrac{1}{a_n}$ for $n = 0, 1, 2, \ldots$.

8. $a_0 = 1$ and $a_{n+1} = 1 - \dfrac{2}{a_n}$ for $n = 0, 1, 2, \ldots$.

9. $a_0 = 1$ and $a_{n+1} = 3 - \dfrac{2}{a_n}$ for $n = 0, 1, 2, \ldots$.

10. $a_0 = 1$ and $a_{n+1} = 4 - \dfrac{2}{a_n}$ for $n = 0, 1, 2, \ldots$.

11. $a_0 = -1$ and $a_{n+1} = \dfrac{1}{2 - a_n}$ for $n = 0, 1, 2, \ldots$.

12. $a_0 = -1$ and $a_{n+1} = \dfrac{2}{3 - a_n}$ for $n = 0, 1, 2, \ldots$.

13. $a_0 = 0$ and $a_{n+1} = \dfrac{a_n + 1}{a_n + 2}$ for $n = 0, 1, 2, \ldots$.

14. $a_0 = 1$ and $a_{n+1} = \dfrac{a_n + 1}{a_n + 2}$ for $n = 0, 1, 2, \ldots$.

15. $a_0 = 1$ and $a_{n+1} = \sqrt{1 - a_n}$ for $n = 0, 1, 2, \ldots$.

16. $a_0 = 1$ and $a_{n+1} = \sqrt{2 - a_n}$ for $n = 0, 1, 2, \ldots$.

17. Prove that the conclusion of Proposition 8.1.5 holds even if the hypothesis $a_n < b_n$ is omitted.

18. $a_0 = 1$ and $a_{n+1} = \sqrt{3 - a_n}$ for $n = 0, 1, 2, \ldots$.

19. $a_0 = 1$ and $a_{n+1} = \dfrac{1}{1 + a_n}$ for $n = 0, 1, 2, \ldots$.

20. $a_0 = -1$ and $a_{n+1} = \dfrac{1}{a_n} - 1$ for $n = 0, 1, 2, \ldots$.

21. $a_0 = 1$ and $a_{n+1} = \dfrac{3 - 2a_n}{3 + 2a_n}$ for $n = 0, 1, 2, \ldots$.

22. $a_0 = 1$ and $a_{n+1} = \dfrac{1}{n + 1} + \dfrac{1}{a_n}$ for $n = 0, 1, 2, \ldots$.

23. $a_0 = c > 0$ and $a_{n+1} = c + \dfrac{1}{a_n}$ for $n = 0, 1, 2, \ldots$.

24. $a_0 = c > 0$ and $a_{n+1} = \dfrac{c + 1/a_n}{n + 1}$ for $n = 0, 1, 2, \ldots$.

25. $a_0 = 0$, $a_{n+1} = a_n^2 + \dfrac{1}{3}$ for $n = 0, 1, 2, \ldots$.

26. $a_0 = 0$, $a_{n+1} = a_n^2 + \dfrac{1}{4}$ for $n = 0, 1, 2, \ldots$.

27. Suppose $a_0 > 0$, $r > 0$ and $a_{n+1} = a_n + \dfrac{r}{a_n}$ for $n = 0, 1, 2, \ldots$.
   Prove that $\{a_n\}$ diverges.

28. Suppose $a_0 = 1$, $r > 0$ and $a_{n+1} = \dfrac{1}{2}\left(a_n + \dfrac{r}{a_n}\right)$ for $n = 0, 1, 2, \ldots$. Prove that $\{a_n\} \to \sqrt{r}$.

29. Suppose $0 \le a \le b$, $a_0 = a$, $b_0 = b$, and $a_{n+1} = \sqrt{a_n b_n}$, $b_{n+1} = \dfrac{a_n + b_n}{2}$. Prove that the two sequences $\{a_n\}$ and $\{b_n\}$ converge to the same limit. (Hint: After you show by induction that the limits exist, substitute them into the equation $a_{n+1} = (a_n + b_n)/2$ to prove that they are equal.)

30. Suppose $a_0 > 0$, $r > 0$ and $a_{n+1} = r\left(a_n + \dfrac{1}{a_n}\right)$ for $n = 0, 1, 2, \ldots$. For which values of r does $\{a_n\}$ converge and for which does it diverge? Find the value of the limit when it exists. Prove your answer.

31. Complete the proof of Theorem 8.1.1 for decreasing sequences.

32. Suppose $0 < \alpha < 1$, $a_{n+1} = (1 - \alpha)a_n + \alpha a_{n-1}$. Prove that $\{a_n\} \to (a_1 + \alpha a_0)/(1 + \alpha)$.

33. Suppose $a_{n+1} = (a_n + b_n)/2$, and $b_{n+1} = \sqrt{a_{n+1} b_n}$, where a_0, b_0 are positive. Prove that $\{a_n\}$ and $\{b_n\}$ are monotone and converge to the same limit.

34. Suppose $a_0, b_0 > 0$, $a_{n+1} = (a_n + b_n)/2$, $a_{n+1} b_{n+1} = a_n b_n$. Prove that $\{a_n\}$ and $\{b_n\}$ are monotone and converge to the same limit.

35. Suppose $a_{n+1} = \sqrt{r + a_n}$, where $r > 0$, $a_0 > 0$. Prove that $\{a_n\}$ is monotone and converges to the positive root of the equation $x^2 - x - r = 0$.

36. Suppose $a_{n+1} = r/(1 + a_n)$, where $r > 0$ and $a_0 > 0$. Prove that $\{a_n\}$ converges to the positive root of the equation $x^2 + x - r = 0$.

37. Suppose $a_{n+1} = (r/a_n) - 1$, where $r > 0$ and $a_0 < 0$. Prove that $\{a_n\}$ converges to the negative root of the equation $x^2 + x - r = 0$.

38. Suppose $0 < a_1 < a_2$, and $a_n = \sqrt{a_{n-1} a_{n-2}}$ for $n = 3, 4, 5, \ldots$. Prove that $\{a_n\}$ converges.

39. Show that there is a function $f(x)$ such that the equation $f(x) = x$ has a real solution, but the sequence $\{a_n\}$ diverges, where $a_0 = 1$ and $a_{n+1} = f(a_n)$ for all n.

40. Suppose every convergent subsequence of the bounded sequence $\{a_n\}$ converges to the same limit A. Prove that $\{a_n\} \to A$.

41. Complete the proof of Proposition 8.1.4.

42. Suppose
$$a_n = 1 + \frac{1}{2^2} + \frac{1}{3^2} + \cdots + \frac{1}{n^2}.$$
Prove that $\{a_n\}$ converges. (Hint: Use the inequality $1/n^2 < 1/(n-1) - 1/n$.)

43. Let $r \geq 2$ be a real number and suppose

$$a_n = 1 + \frac{1}{2^r} + \frac{1}{3^r} + \cdots + \frac{1}{n^r}.$$

Prove that $\{a_n\}$ converges.

44. Prove that if each d_k is a digit $0, 1, \ldots$, or 9, then

$$\frac{d_1}{10} + \frac{d_2}{10^2} + \frac{d_3}{10^3} + \cdots + \frac{d_n}{10^n} < 1.$$

Conclude that the sequence $\{a_n\}$ where $a_{n+1} = a_n + (d_n/10^n)$ converges.

45. Does the sequence defined by $a_0 = 1$ and $a_{n+1} = a_n + a_n^{-n}$ converge or diverge? Justify your answer.

8.2   THE CAUCHY PROPERTY

The convergence proofs of Chapter 7 all made use of the proposed limit. Frequently, however, it is necessary to prove the convergence of a sequence whose limit is unknown. In such cases an intrinsic convergence criterion is handy. Cauchy provided such a criterion in the context of series, but, for pedagogical and other reasons, it has become customary to introduce his ideas in the simpler context of sequences first.

A sequence is said to have the *Cauchy Property* if for every $\varepsilon > 0$ there exists an index n' such that

$$|a_{n+m} - a_n| < \varepsilon \qquad \text{for all } n \geq n' \text{ and } m = 1, 2, 3, \ldots$$

or, equivalently,

$$\{a_{n+m}\}_{m=0}^{\infty} \subset (a_n - \varepsilon, a_n + \varepsilon) \qquad \text{for all } n \geq n'.$$

In other words, given any $\varepsilon > 0$, there is an index n' such that for every $n \geq n'$ the ε-neighborhood of a_n contains all the subsequent terms of the sequence. As will be seen below, the Cauchy Property is synonymous with convergence. The great advantage of this new formulation of convergence is that it makes it possible to prove the convergence of a sequence even when the limit is unknown.

Example 8.2.1 If $\{a_n\}$ is any sequence such that $|a_{n+1} - a_n| \leq 1/7^n$ then it has the Cauchy Property. To see this let $\varepsilon > 0$ and let n' be an index such that $1/(6 \cdot 7^{n-1}) < \varepsilon$ for all $n \geq n'$ (n' exists because $1/7^n \to 0$). Then, for all $n \leq n'$ we have

$$|a_{n+m} - a_n| \leq |a_{n+m} - a_{n+m-1} + a_{n+m-1} - a_{n+m-2} + \cdots + a_{n+1} - a_n|$$

$$\leq |a_{n+m} - a_{n+m-1}| + |a_{n+m-1} - a_{n+m-2}| + \cdots + |a_{n+1} - a_n|$$

$$\leq \frac{1}{7^{n+m-1}} + \frac{1}{7^{n+m-2}} + \cdots + \frac{1}{7^n}$$

$$\leq \frac{1}{7^n}\left(\frac{1}{7^{m-1}} + \frac{1}{7^{m-2}} + \cdots + 1\right) = \frac{1}{7^n} \cdot \frac{1 - \frac{1}{7^m}}{1 - \frac{1}{7}}$$

$$< \frac{1}{7^n} \cdot \frac{1}{1 - \frac{1}{7}} = \frac{1}{6 \cdot 7^{n-1}} < \varepsilon.$$

Before the equivalence of the Cauchy Property with convergence is proved, it is necessary to establish two lemmas. Both the statement and the proof of the first of these resemble those of Lemma 7.2.2.

Lemma 8.2.2 *If a sequence has the Cauchy Property then it is bounded.*

Proof. Since $\{a_n\}$ has the Cauchy Property, it follows that there exists an index n' such that

$$\{a_{n'+m}\}_{m=0}^{\infty} \subset (a_{n'} - 1, a_{n'} + 1).$$

Since

$$\{a_n\}_{n=0}^{\infty} = \{a_0, a_1, a_2, \cdots, a_{n'-1}\} \cup \{a_{n'+m}\}_{m=0}^{\infty}$$

it follows from Lemma 6.2.4 that the given sequence is bounded.   □

Lemma 8.2.3 *If the sequence $\{a_n\}$ has the Cauchy Property and the subsequence $\{a_{n_k}\}$ converges to A, then $\{a_n\}$ also converges to A.*

Proof. Let $\varepsilon > 0$. Since $\{a_{n_k}\} \to A$, there exists an index k' such that

$$|a_{n_k} - A| < \frac{\varepsilon}{2} \qquad \text{for all } k \geq k'.$$

Since the sequence $\{a_n\}$ has the Cauchy Property, there exists an index k'' such that

$$|a_{n_k+m} - a_{n_k}| < \frac{\varepsilon}{2} \qquad \text{for all } k \geq k'' \text{ and } m = 1, 2, 3, \ldots$$

or equivalently,

$$|a_n - a_{n_k}| < \frac{\varepsilon}{2} \qquad \text{for all } k \geq k'' \text{ and } n \geq n_k.$$

Hence, if we set $k^* = \max\{k', k''\}$, then for all $n \geq n_{k^*}$

$$|a_n - A| = |(a_n - a_{n_{k^*}}) + (a_{n_{k^*}} - A)| \leq |a_n - a_{n_{k^*}}| + |a_{n_{k^*}} - A|$$

$$< \frac{\varepsilon}{2} + \frac{\varepsilon}{2} = \varepsilon.$$

Hence $\{a_n\} \to A$.   □

Loosely speaking, the above argument can be paraphrased as follows (see Figure 8.2). Because $\{a_{n_k}\} \to A$, for sufficiently large k, a_{n_k} lies within $\varepsilon/2$ of A. Because $\{a_n\}$ has the Cauchy Property, for sufficiently large k, a_n lies within $\varepsilon/2$ of a_{n_k} whenever $n \geq n_k$. Consequently, for sufficiently large k, a_n lies within $\varepsilon = \varepsilon/2 + \varepsilon/2$ of A whenever $n \geq n_k$. Thus, $\{a_n\} \to A$.

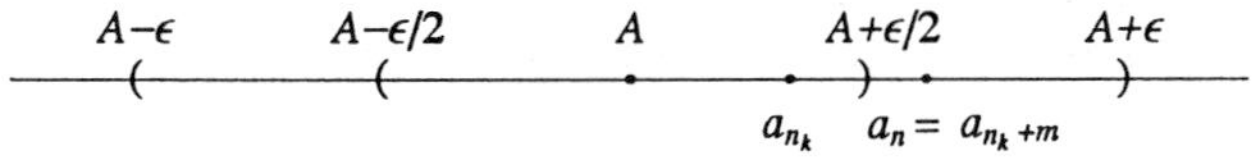

Fig. 8.2   A proof of convergence

Theorem 8.2.4 *A sequence has the Cauchy Property if and only if it is a convergent sequence.*

Proof. Suppose $\{a_n\}$ is a convergent sequence and A is its limit. Let $\varepsilon > 0$ be given and let n' be an index such that

$$|a_n - A| < \frac{\varepsilon}{2} \qquad \text{for all } n \geq n'.$$

If m is any positive integer, then $n + m > n'$ for all $n \geq n'$ and so

$$|a_{n+m} - A| < \frac{\varepsilon}{2} \qquad \text{for all } n \geq n' \text{ and } m = 1, 2, 3, \ldots.$$

Consequently,

$$|a_{n+m} - a_n| = |(a_{n+m} - A) - (a_n - A)| \leq |a_{n+m} - A| + |a_n - A|$$
$$< \frac{\varepsilon}{2} + \frac{\varepsilon}{2} = \varepsilon \qquad \text{for all } n \geq n' \text{ and } m = 1, 2, 3, \ldots.$$

Thus, the convergent sequence $\{a_n\}$ has the Cauchy Property.

Conversely, suppose the sequence $\{a_n\}$ has the Cauchy Property. It follows from Lemma 8.2.2 that $\{a_n\}$ is bounded. Since $\{a_n\}$ is bounded it has a convergent subsequence and so, by Lemma 8.2.3, $\{a_n\}$ is convergent.   □

Alternately, the first part of the above proof could be rephrased as follows (see Figure 8.3). Since $\{a_n\} \to A$ there exists n' such that

$$\{a_n\}_{n=n'}^{\infty} \subset (A - \varepsilon/2, A + \varepsilon/2)$$

and consequently

$$|a_{n+m} - a_n| < \varepsilon \qquad \text{for all } n \geq n' \text{ and all } m.$$

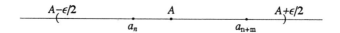

Fig. 8.3   The Cauchy Property is equivalent to convergence

Exercises 8.2

1. Suppose $\{a_n\}$ is a sequence such that $|a_{n+1} - a_n| < 2^{-n}$ for all $n > 0$. Prove that $\{a_n\}$ has the Cauchy Property and is therefore convergent.

2. Suppose

$$a_n = 1 - \frac{1}{2^2} + \frac{1}{3^3} - \frac{1}{4^4} + \cdots + \frac{(-1)^{n-1}}{n^n}.$$

Prove that the sequence $\{a_n\}$ has the Cauchy Property and therefore converges.

3. Suppose $\{a_n\}$ is a sequence such that $|a_{n+1} - a_n| < n^{-2}$ for all $n > 0$. Prove that $\{a_n\}$ has the Cauchy Property and is therefore convergent. (Hint: Use the inequality $1/n^2 < 1/(n-1) - 1/n$.)

4. Suppose the sequence $\{a_n\}$ has the following property: For every $\varepsilon > 0$ there exists an interval $(a, a + \varepsilon)$ that contains almost all the terms of $\{a_n\}$. Prove that the sequence $\{a_n\}$ has the Cauchy Property.

5. Suppose the sequences $\{a_n\}$ and $\{b_n\}$ both have the Cauchy Property. Prove directly (without recourse to Theorem 8.2.4) that the following sequences also have the Cauchy Property.
   a) $\{a_n + b_n\}$ b) $\{a_n - b_n\}$ c) $\{ra_n\}$
   d) $\{a_n b_n\}$, provided $a_n, b_n \geq 0$
   e) $\{1/a_n\}$ provided there is a number $a > 0$ such that $|a_n| > a$ for all n.

6. Find a sequence $\{a_n\}$ of non-zero terms that has the Cauchy Property but for which the sequence $\{1/a_n\}$ does not have the Cauchy Property.

7. Suppose $0 < c < 1$. Prove that if the sequence $\{a_n\}$ has the property that $|a_{n+2} - a_{n+1}| < c|a_{n+1} - a_n|$ for all $n = 0, 1, 2, \ldots$, then the sequence $\{a_n\}$ has the Cauchy Property.

Chapter Summary

It is proved on the basis of the Completeness Axiom that bounded monotone sequences necessarily converge. While this is only an existential assertion, it can be very useful in the actual evaluation of limits. Moreover, as another proposition asserts that every sequence contains a monotone subsequence, this implies the important Bolzano–Weierstrass Theorem: *Every bounded sequence contains a convergent subsequence.* Next, these preliminaries are used to provide a powerful tool, the Cauchy Property, for recognizing the convergence of sequences even when their limits are unknown.

9

The Convergence of Infinite Series

The tools constructed in Chapters 7 and 8 are now applied to a variety of infinite series. Several criteria guaranteeing convergence and divergence of series are offered.

9.1 STOCK SERIES

To every sequence $\{a_n\}$ we associate a new sequence $\{d_n\}$ of *partial sums* where for $n = 0, 1, 2, \ldots,$

$$d_n = a_0 + a_1 + a_2 + \cdots + a_n. \tag{1}$$

If, for example, $\{a_n\} = \{1/2^n\}$, then the associated sequence of partial sums is

$$d_0 = a_0 = 1$$

$$d_1 = a_0 + a_1 = 1 + \frac{1}{2} = 1\frac{1}{2}$$

$$d_2 = a_0 + a_1 + a_2 = 1 + \frac{1}{2} + \frac{1}{4} = 1\frac{3}{4}$$

$$d_3 = a_0 + a_1 + a_2 + a_3 = 1 + \frac{1}{2} + \frac{1}{4} + \frac{1}{8} = 1\frac{7}{8}$$

$$d_4 = a_0 + a_1 + a_2 + a_3 + a_4 = 1 + \frac{1}{2} + \frac{1}{4} + \frac{1}{8} + \frac{1}{16} = 1\frac{15}{16}$$

and in general, by Proposition 1.1.2,

$$d_n = 1 + \frac{1}{2} + \frac{1}{4} + \cdots + \frac{1}{2^n} = \frac{1 - \left(\frac{1}{2}\right)^{n+1}}{1 - \frac{1}{2}} = 2 - \left(\frac{1}{2}\right)^n.$$

Similarly, if $\{a_n\} = \{n\}$, then

$$
\begin{aligned}
d_0 &= a_0 = 0 \\
d_1 &= a_0 + a_1 = 0 + 1 = 1 \\
d_2 &= a_0 + a_1 + a_2 = 0 + 1 + 2 = 3 \\
d_3 &= a_0 + a_1 + a_2 + a_3 = 0 + 1 + 2 + 3 = 6 \\
d_4 &= a_0 + a_1 + a_2 + a_3 + a_4 = 0 + 1 + 2 + 3 + 4 = 10
\end{aligned}
$$

and in general, by Exercise 10,

$$d_n = 0 + 1 + 2 + \cdots + n = \frac{n(n+1)}{2}.$$

If $\{a_n\} = \{1\}$, then $\{d_n\} = \{n+1\}$ and if $\{a_n\} = \{(-1)^n\}$ then $\{d_n\} = 1, 0, 1, 0, 1, 0, \ldots = \{(1 + (-1)^n)/2\}$. If the associated sequence $\{d_n\}$ happens to be convergent, with limit, say, A, we write

$$\sum a_n = a_0 + a_1 + a_2 + a_3 + \cdots = A \tag{2}$$

and say that the *infinite series* $\sum a_n$ *converges*. If $\{d_n\}$ diverges, $\sum a_n$ is said to be *divergent*. Thus, since it was shown above that for the series $\sum 1/2^n$ we have $d_n = 2 - 1/2^n$, and since $\{2 - 1/2^n\} \to 2$, it follows that

$$\sum \frac{1}{2^n} = 2.$$

On the other hand, since $\{n(n+1)/2\}$ is a divergent sequence, the infinite series $\sum n$ diverges.

The above convergent series is atypical. Most convergent series do not have integer or even rational sums and it is in general impossible to describe their limits in terms of simple closed expressions. In general one makes do by deciding on theoretical grounds whether a series converges and then, if it does and if the numerical value of its sum happens to be of interest, one can use the partial sums of the series as estimates of this total sum. In Propositions 9.1.1, 9.1.2, and 9.1.5 the convergence and divergence of some stock series are established. The subsequent Propositions 9.1.3, 9.2.1, and 9.3.3 show how this information can be used to determine the nature of many other series.

Proposition 9.1.1 $\sum a^n = 1/(1-a)$ *if* $|a| < 1$.

Proof. By Proposition 1.1.2

$$d_n = 1 + a + a^2 + a^3 + \cdots + a^n = \frac{1 - a^{n+1}}{1 - a}.$$

By Proposition 7.2.7 $\{a^{n+1}\} \to 0$ if $|a| < 1$ and so

$$\sum a_n = \lim_{n \to \infty} d_n = \lim_{n \to \infty} \frac{1 - a^{n+1}}{1 - a} = \frac{1}{1 - a}. \qquad \Box$$

Since the proof of the following proposition employs a definite integral, it is not quite proper in this context. After all, the definite integral has yet to be defined here and both its definition and the Fundamental Theorem of Calculus that is used to evaluate the integral are non-trivial matters. Still, since the integral method for proving the convergence of infinite series is such a useful and elegant tool, it is included. An alternate method is outlined in Exercise 13.

Proposition 9.1.2 $\sum 1/n^p$ *converges if and only if* $p > 1$.

Proof. The partial sums

$$d_n = \frac{1}{1^p} + \frac{1}{2^p} + \frac{1}{3^p} + \cdots + \frac{1}{n^p}$$

are increasing since

$$d_{n+1} = d_n + \frac{1}{(n+1)^p} > d_n.$$

Suppose first that $p > 1$. To prove that these partial sums are also bounded examine the graph of $f(x) = 1/x^p$ for $x > 0$. Since this function is decreasing, the relative position of this graph to the rectangles depicted in Fig. 9.1 is

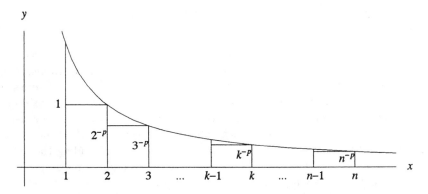

Fig. 9.1   A proof of the convergence of a series.

accurate. The rectangle between $x = k - 1$ and $x = k$ has height $1/k^p$ and width 1, and so it has area $1/k^p$. On the other hand, the region between the graph of $f(x)$ and the x axis from $x = 1$ to $x = n$ has area $\int_1^n dx/x^p$. It therefore follows that

$$d_n < \frac{1}{1^p} + \int_1^n \frac{dx}{x^p} = 1 + \left[\frac{-1}{(p-1)x^{p-1}} \right]_{x=1}^{x=n}$$

$$= 1 - \frac{1}{(p-1)n^{p-1}} + \frac{1}{p-1} < 1 + \frac{1}{p-1} = \frac{p}{p-1}.$$

Thus, the increasing sequence $\{d_n\}$ is bounded above by $p/(p-1)$ and so $\sum 1/n^p$ converges.

If $p = 1$ and $e_n = 1 + 1/2 + 1/3 + \cdots + 1/n$ then for every positive integer m,

$$e_{2^m} = \frac{1}{1} + \frac{1}{2} + \cdots + \frac{1}{2^m}$$

$$= \left(\frac{1}{1} + \frac{1}{2} + \cdots + \frac{1}{2^{m-1}} \right) + \left(\frac{1}{2^{m-1}+1} + \frac{1}{2^{m-1}+2} + \cdots + \frac{1}{2^m} \right)$$

$$= e_{2^{m-1}} + \left(\frac{1}{2^{m-1}+1} + \frac{1}{2^{m-1}+2} + \cdots + \frac{1}{2^{m-1}+2^{m-1}} \right)$$

$$> e_{2^{m-1}} + 2^{m-1} \left(\frac{1}{2^m} \right) = e_{2^{m-1}} + \frac{1}{2}.$$

It follows by an easy inductive argument that $e_{2^m} > m/2$ and hence the sequence $\{e_n\}$ is unbounded. By Lemma 7.2.2 it may be concluded that $\{e_n\}$ and $\sum 1/n$ diverge. Finally, if $p < 1$ and

$$f_n = 1 + \frac{1}{2^p} + \frac{1}{3^p} + \cdots + \frac{1}{n^p} > 1 + \frac{1}{2} + \frac{1}{3} + \cdots + \frac{1}{n} = e_n$$

then the unboundedness and divergence of $\{f_n\}$ follows from the unboundedness of $\{e_n\}$. $\Box$

The series $\sum 1/n$ is called the *harmonic series*. Its divergence was first demonstrated circa 1350 by Nicole Oresme (1323–1382) whose argument was similar to that used in the proof of the above proposition. This series is notoriously slow to diverge and this turns it into the perfect counterexample to the misconception that "All that we really need to do to see whether a series converges is to check it with a calculator." For, if at the time of the "Big Bang" (20 billion years ago), a calculator had started adding the terms of this series at the rate of one million additions per second, its registers would have accumulated only a partial sum of 55 thus far.

As mentioned in Chapter 5 and demonstrated in the next chapter, the actual value of $\sum 1/n^p$ for integer values of p is of considerable interest to mathematicians. At this point, however, we are only interested in the issue of whether a given series converges or not. Consequently, the value of the sum, even when the series does converge, is of no concern to us. A discussion of the general question of the evaluation of limits of infinite series will be found in the next section.

The proof of the following proposition follows immediately from the definition of the convergence of series and is therefore relegated to Exercises 4–6.

Proposition 9.1.3 *Let $r \neq 0$. Then*

1. $\sum a_n$ *converges if and only if $\sum r a_n$ converges, in which case we also have $\sum r a_n = r \sum a_n$.*

2. *If $\sum a_n$ and $\sum b_n$ both converge so does $\sum (a_n + b_n)$ and $\sum (a_n + b_n) = \sum a_n + \sum b_n$.*

3. *If $\sum a_n$ converges, then $\{a_n\} \to 0$.*

Example 9.1.4 The series $\sum \cos nx$ diverges whenever x is a rational multiple of π. To see this, suppose $x = (s/t)\pi$ where s and t are integers. If n is any integer that is divisible by t, then nx is an integer multiple of π and so $\cos nx = \pm 1$ and clearly $\{\cos nx\}$ does not converge to 0. Hence, by the third part of the above proposition, $\sum \cos nx$ diverges for these values of x. It can also be shown that the given series diverges for all values of x (Exercise 3).

The next corollary is an immediate consequence of Proposition 9.1.3.3 and Proposition 7.2.7.

Proposition 9.1.5 *If $|a| \geq 1$ then $\sum a^n$ diverges.*

Exercises 9.1

1. Evaluate the following infinite series.

a) $\sum \left(\frac{2}{3}\right)^n$

b) $\sum \left(-\frac{3}{4}\right)^n$

c) $\sum \left(\frac{3}{11}\right)^n$

d) $\sum \left(-\frac{7}{9}\right)^n$

e) $\sum \left[\left(\frac{1}{2}\right)^n + \left(\frac{1}{3}\right)^n\right]$

f) $\sum \left[5 \left(\frac{1}{2}\right)^n + 12 \left(\frac{1}{3}\right)^n\right]$

g) $\sum \left[5 \left(-\frac{2}{3}\right)^n + 12 \left(\frac{3}{4}\right)^n\right]$

2. Which of the following series converge and which diverge? Justify your answers.

a) $\sum n^{3/2}$

b) $\sum n^{-3/2}$

c) $\sum n^{-2/3}$

d) $\sum \left(5n^{-3/4} + 2n^{3/2}\right)$

e) $\sum (3^n - 2^n)$

f) $\sum \left(\dfrac{5}{2^n} - \dfrac{4}{3^n}\right)$

g) $\sum \left(\dfrac{1}{n^3} + \dfrac{1}{3^n}\right)$

h) $\sum (-1)^n$

3. Show that if x/π is an irrational number then $\sum \cos nx$ diverges.

4. Prove Proposition 9.1.3.2. (Hint: Set $d_n = \sum_{k=0}^n a_n$ and $e_n = \sum_{k=0}^n b_n$ and use Theorem 7.2.3).

5. Prove Proposition 9.1.3.1. (See hint for Exercise 4.)

6. Prove Proposition 9.1.3.3. (Use the fact that in the hint for Exercise 5 $a_n = d_n - d_{n-1}$.)

7. Prove that $\sum_{n=2}^\infty 1/(n(\ln n)^p)$ converges when $p > 1$ and diverges when $p \le 1$.

8. For which rational values of r does $\sum \sin n(r\pi)$ converge and for which does it diverge?

9. Show that if x/π is an irrational number then $\sum \sin nx$ diverges.

10. Prove that $1 + 2 + 3 + \cdots + n = n(n+1)/2$ for $n = 0, 1, 2, \ldots$.

11. Suppose $a_0 = 1$ and $a_{n+1} = a_n + 1/(na_n)$ for $n = 0, 1, 2, \ldots$. Does $\{a_n\}$ converge or diverge?

12. Suppose $a_0 = 1$ and $a_{n+1} = a_n + 1/(n^2 a_n)$ for $n = 0, 1, 2, \ldots$. Does $\{a_n\}$ converge or diverge?

13. Let $p > 1$ and $d_n = 1 + \dfrac{1}{2^p} + \dfrac{1}{3^p} + \cdots + \dfrac{1}{n^p}$. Prove that

a) $d_{2^m} < d_{2^{m-1}} + \left(\dfrac{1}{2^{p-1}}\right)^m$ for $m = 1, 2, 3, \ldots$

b) $d_{2^m} < \dfrac{1}{1 - 1/2^{p-1}}$ for $m = 0, 1, 2, \ldots$

c) $\sum 1/n^p$ converges.

14. Describe the main mathematical achievements of Nicole Oresme.

9.2   SERIES OF POSITIVE TERMS

At this point it is necessary to restrict the discussion to *series of positive terms*, by which we mean series all of whose terms are non-negative. The discussion of the general case will be resumed in the next section. The following proposition is commonly called the *comparison test*.

Proposition 9.2.1 *Suppose $\sum a_n$ and $\sum b_n$ are series such that*

$$0 \le a_n \le b_n. \tag{3}$$

Then,

$$\text{if } \sum b_n \quad converges \text{ so does } \sum a_n; \tag{4}$$
$$\text{if } \sum a_n \quad diverges \text{ so does } \sum b_n. \tag{5}$$

Proof. Set

$$d_n = a_0 + a_1 + a_2 + \cdots + a_n \qquad n = 0, 1, 2, \ldots$$
$$e_n = b_0 + b_1 + b_2 + \cdots + b_n \qquad n = 0, 1, 2, \ldots$$

It follows from Inequality (3) that $\{d_n\}$ and $\{e_n\}$ are increasing sequences. Hence, by the Monotone Convergence Theorem, each of them converges or diverges according as it is bounded or not. Inasmuch as it also follows from Inequality (3) that

$$0 \le d_n \le e_n$$

we conclude that

$$\text{if } \{e_n\} \text{ is bounded (convergent), so is } \{d_n\}, \tag{6}$$
$$\text{if } \{d_n\} \text{ is unbounded (divergent), so is } \{e_n\}. \tag{7}$$

By the definition of convergence of series Statements (6) and (7) are equivalent to Statements (4) and (5) respectively. □

Example 9.2.2 Determine the convergence or divergence of $\sum 1/(\sqrt{n}3^n)$ and $\sum (2n+1)/(3n^2 + n + 1)$.

Note that $1/(\sqrt{n}3^n) < 1/3^n$ and, by Proposition 9.1.1, $\sum 1/3^n$ converges. Hence it follows from Proposition 9.2.1.2 that $\sum 1/(\sqrt{n}3^n)$ also converges. Turning to the second given series, observe that the inequality

$$\frac{2n+1}{3n^2 + n + 1} > \frac{1}{2n}$$

is established by cross multiplication. Since, by Propositions 9.1.2 and 9.1.3 $\sum 1/(2n)$ is divergent, it follows from Proposition 9.2.1.2 that

$$\sum (2n+1)/(3n^2 + n + 1)$$

is also divergent.

While Proposition 9.2.1 is helpful in converting questions of convergence of new series to those of old series, the proposition below provides an internal convergence test. It was first used by D'Alembert in 1768 to discuss the convergence of the binomial series and subsequently extended to all series by Cauchy in 1821.

Proposition 9.2.3 (Ratio test) *Suppose $\sum a_n$ is a series such that $a_n > 0$ for all n, and suppose*

$$\lim_{n \to \infty} \frac{a_{n+1}}{a_n} = q.$$

Then $q \geq 0$, and

$$\sum a_n \text{ converges if } q < 1;$$
$$\sum a_n \text{ diverges if } q > 1.$$

Proof. That $q \geq 0$ follows from the fact that $a_{n+1}/a_n > 0$. Let (q'', q') be any neighborhood of q that does not contain either of the numbers ± 1 (see Fig. 9.2). By the definition of convergence of sequences, the interval

Fig. 9.2   The relative position of q, 1, and (q'', q')

(q'', q') contains almost all the terms of $\{a_{n+1}/a_n\}$. Delete all the terms of the original sequence $\{a_n\}$ for which $a_{n+1}/a_n \notin (q'', q')$. By Proposition 7.2.13 this does not affect the convergence or divergence of $\sum a_n$ and so we may assume without loss of generality that

$$q'' < \frac{a_{n+1}}{a_n} < q' \qquad \text{for all } n = 0, 1, 2, \dots .$$

An easy inductive argument allows us to conclude that

$$a_0 q''^n < a_n < a_0 q'^n \qquad \text{for all } n = 1, 2, 3, \dots .$$

If $q < 1$, then $q' < 1$ because (q'', q') contains q but does not contain 1. Hence $\sum a_n$ converges by comparison with the convergent infinite geometric progression $a_0 \sum q'^n$.

If $q > 1$, then $q'' > 1$ and so $\sum a_n$ diverges by comparison with the divergent infinite geometric progression $a_0 \sum q''^n$. □

Example 9.2.4 The series $\sum n^2/3^n$ converges because

$$\frac{a_{n+1}}{a_n} = \frac{(n+1)^2/3^{n+1}}{n^2/3^n} = \left(\frac{n+1}{n}\right)^2 \cdot \frac{1}{3} \to 1^2 \cdot \frac{1}{3} < 1.$$

On the other hand, the series $\sum 2^n/n^5$ diverges because

$$\frac{a_{n+1}}{a_n} = \frac{2^{n+1}/(n+1)^5}{2^n/n^5} = 2\left(\frac{n}{n+1}\right)^5 \to 2 > 1.$$

The case $q = 1$, whose discussion was omitted from Proposition 9.2.3, is ambiguous. For both of the series $\sum 1/n$ and $\sum 1/n^2$ we have

$$\frac{a_{n+1}}{a_n} \to 1$$

and yet, by Proposition 9.1.2, the first of these diverges whereas the second converges.

Exercises 9.2

1. Decide whether the following series converge or diverge. Justify your answers.

a) $\displaystyle\sum \frac{n}{n+2}$

b) $\displaystyle\sum \frac{1}{2n+3}$

c) $\displaystyle\sum \frac{1}{n^2+1}$

d) $\displaystyle\sum \frac{n!}{n^n}$

e) $\displaystyle\sum_{n=2}^{\infty} \frac{1}{(\ln n)^{\ln n}}$

f) $\displaystyle\sum \frac{1}{\sqrt{n}}$

g) $\displaystyle\sum \frac{n^2+1}{3n^3-n}$

h) $\displaystyle\sum \frac{n^2+1}{3n^4-n}$

i) $\displaystyle\sum \frac{n}{3^n}$

j) $\displaystyle\sum \frac{3}{n^2+1}$

k) $\displaystyle\sum \left(\frac{1}{\sqrt{n}} - \frac{1}{n^2}\right)$

l) $\displaystyle\sum \left(\frac{1}{\sqrt{n}} - \frac{1}{\sqrt{n+1}}\right)$

m) $\displaystyle\sum \left(\frac{1}{\sqrt{n}} + \frac{1}{\sqrt{n+1}}\right)$

n) $\displaystyle\sum \left(\frac{1}{n} - \frac{1}{n+1}\right)$

o) $\displaystyle\sum \left(\frac{1}{n} + \frac{1}{n+1}\right)$

p) $\displaystyle\sum \left(\frac{1}{n^2} - \frac{1}{(n+1)^2}\right)$

q) $\displaystyle\sum \left(\frac{1}{n^2} + \frac{1}{(n+1)^2}\right)$

2. Suppose $a_n \geq 0$ for $n = 0, 1, 2, \ldots$ and $\sqrt[n]{a_n} \to q$. Then $\sum a_n$ converges if $q < 1$ and diverges if $q > 1$. (This is Cauchy's *Root Test*.)

3. Suppose $a_n \geq 0$ for $n = 0, 1, 2, \ldots$, and $\sum a_n$ converges. Prove that $\sum a_n^2$ also converges.

4. Suppose $\sum a_n$ and $\sum b_n$ both converge. Prove that if $a_n, b_n \geq 0$, then $\sum a_n b_n$ also converges.

5. Let $\{a_n\}$ be the sequence obtained by deleting from $\mathbb{N} = \{n\} = 1, 2, 3, \ldots$ all those integers whose decimal expansion contains the digit 1, that is

$$\{a_n\} = 2, 3, \ldots, 9, 20, 22, 23, 24, \ldots, 30, 32, 33, \ldots.$$

Prove that the series $\sum 1/a_n$ converges.

6. Let $\{b_n\}$ be the sequence obtained by deleting from $\mathbb{N}$ all the integers whose decimal expansion does not contain the digit 1, that is

$$\{b_n\} = 1, 10, 11, 12, 13, \ldots, 19, 21, 31, \ldots.$$

Does the series $\sum 1/b_n$ converge or diverge? Jusify your answer.

7. For any fixed α, β, γ, the *Hypergeometric Series* is the series

$$1 + \sum_{n=1}^{\infty} \frac{\alpha(\alpha+1)\ldots(\alpha+n-1)\beta(\beta+1)\cdots(\beta+n-1)}{n!\gamma(\gamma+1)\cdots(\gamma+n-1)} x^n.$$

Prove the following:

a) The Hypergeometric Series converges for $0 \le x < 1$.

b) The Hypergeometric Series diverges for $1 < x$.

c) The Hypergeometric Series converges for $x = 1$ if $\gamma > \alpha + \beta$.

d) The Hypergeometric series diverges for $x = 1$ if $\gamma \le \alpha + \beta$.

8. Suppose $\sum a_n$ and $\sum b_n$ are convergent series such that $a_n \ge 0$ and $b_n \ge 0$ for all n. If $c_n = \max\{a_n, b_n\}$, does the series $\sum c_n$ necessarily converge? Justify your answer.

9. Let $\sum a_n = A$ and $\sum b_n = B$ be convergent series of positive terms and suppose $c_n = \sum_{k=0}^{n} a_{n-k} b_k$ for each $n = 0, 1, 2, \ldots$. Prove that $\sum c_n = AB$. Interpret this as a *multiplication of series*.

10. Let $\sum a_n$ be a series of positive terms. Prove that $\sum a_n$ converges if and only if $\sum 2^m a_{2^m}$ converges.

11. Suppose $\sum a_n$ and $\sum b_n$ are both divergent series of positive terms. Does $\sum (a_n + b_n)$ necessarily diverge? Justify your answer.

9.3 SERIES OF ARBITRARY TERMS

We resume the discussion of series of arbitrary terms. The treatment of this more general case, wherein the terms are not necessarily positive, calls for a deeper examination of the notion of convergence. For the purposes of this discussion fix the series $\sum a_n$ and recall that it is convergent if the sequence of its partial sums $\{a_0 + a_1 + \cdots + a_n\}$ converges. Informally speaking, this is tantamount to the expectation that the tails

$$a_{n+1} + a_{n+2} + a_{n+3} + \cdots \qquad (8)$$

converge to 0 as n goes to infinity; in other words that

$$\lim_{n \to \infty} (a_{n+1} + a_{n+2} + a_{n+3} + \cdots) = 0. \qquad (9)$$

At first glance this may not seem a very helpful observation since we have simply replaced the given infinite series $\sum a_n$ by a sequence of infinite series (Equation 9). The advantage lies in the fact that whereas the value of $\sum a_n$ is generally unknown, even when it does converge, Equation (9) involves a known limit, namely 0. The convergence of a sequence to 0 is tantamount to its terms being arbitrarily small. Consequently, to verify Equation (9) it suffices to show that for every $\varepsilon > 0$ and for sufficiently large n,

$$|a_{n+1} + a_{n+2} + a_{n+3} + \cdots| < \varepsilon. \qquad (10)$$

It may seem that we have merely replaced one infinite series by another, but in fact progress has been made. To verify Inequality (10) it suffices to show that for each of the partial tails $a_{n+1} + a_{n+2} + \cdots + a_{n+m}$ we have

$$|a_{n+1} + a_{n+2} + \cdots + a_{n+m}| < \varepsilon'$$

where ε' is any positive number less than ε, e.g., $\varepsilon/2$. Thus the proof of convergence of an infinite series has been reduced to the verification of an inequality involving only a finite number of terms. This discussion is now formalized by means of a definition and a proposition.

A series $\sum a_n$ is said to have the *Cauchy Property* if and only for each $\varepsilon > 0$ there exists an index n' such that

$$|a_{n+1} + a_{n+2} + \cdots + a_{n+m}| < \varepsilon \qquad \text{for all } n \geq n' \text{ and } m = 1, 2, 3, \ldots$$

Cauchy took it for granted that this property is actually equivalent to convergence. Modern standards of rigor, however, require a proof. The details of the proof have already been carried out in the guise of Theorem 8.2.4. The reason the proof of this important principle was placed in the previous section is that, despite the fact that it was first formulated for series, its proof is considerably simplified by the use of the notion of a sequence.

Theorem 9.3.1 *The series $\sum a_n$ converges if and only if it has the Cauchy Property.*

Proof. As usual, let $d_n = a_0 + a_1 + \cdots + a_n$. Since

$$d_{n+m} - d_n = a_{n+1} + a_{n+2} + \cdots + a_{n+m}$$

it follows that the series $\sum a_n$ has the Cauchy Property if and only if the sequence $\{d_n\}$ has the Cauchy Property. The statement of the theorem now follows from Theorem 8.2.4. $\qquad\qquad\square$

It follows from Exercise 8.2.2 that the series $\sum (-1)^{n-1}/n^n$ converges. The first seven partial sums $1, .75, .787037, .783131, .783451, .783429, .783431$ gives us a fairly good idea of the value of the limit. We are now ready to formulate a very general and eminently reasonable criterion for the convergence of infinite series with terms that are not necessarily positive.

Proposition 9.3.2 *If $\sum |a_n|$ converges then so does $\sum a_n$. Moreover $|\sum a_n| \leq \sum |a_n|$.*

Proof. Let $\sum a_n$ be a series such that $\sum |a_n|$ converges. We shall show that $\sum a_n$ has the Cauchy Property. Let $\varepsilon > 0$. Since $\sum |a_n|$ is convergent it has the Cauchy Property and so there exists an index n' such that

$$\big||a_{n+1}| + |a_{n+2}| + \cdots + |a_{n+m}|\big| < \varepsilon \qquad \text{for all } n \geq n' \text{ and } m = 0, 1, 2, \ldots .$$

However,

$$|a_{n+1} + a_{n+2} + \cdots + a_{n+m}| \leq |a_{n+1}| + |a_{n+2}| + \cdots + |a_{n+m}|$$
$$= \big||a_{n+1}| + |a_{n+2}| + \cdots + |a_{n+m}|\big| < \varepsilon$$

for the same values of n and m. Hence $\sum a_n$ has the Cauchy Property and so is convergent. Moreover, since $-|a_n| \leq a_n \leq |a_n|$, it follows that

$$-(|a_0| + |a_1| + |a_2| + \cdots + |a_n|) \leq a_0 + a_1 + a_2 + \cdots + a_n$$
$$\leq |a_0| + |a_1| + |a_2| + \cdots + |a_n|.$$

Hence, by Proposition 7.2.10

$$-\sum |a_n| \leq \sum a_n \leq \sum |a_n|$$

or

$$\left|\sum a_n\right| \leq \sum |a_n| \qquad\qquad\square$$

Example 9.3.3 By Proposition 9.1.2 the series $\sum 1/n^3$ converges. It follows from Proposition 9.3.2 that the series

$$1 + \frac{1}{2^3} - \frac{1}{3^3} + \frac{1}{4^3} + \frac{1}{5^3} - \frac{1}{6^3} + \frac{1}{7^3} + \frac{1}{8^3} - \frac{1}{9^3} + \cdots$$

also converges.

Example 9.3.4 Since it is known by Proposition 9.1.1 that $\sum 1/2^n$ converges, and since $|\cos \alpha| \leq 1$, it follows from Proposition 9.3.2 that

$$\sum \frac{\cos nx}{2^n}$$

converges for every value of x.

Corollary 9.3.5 *Suppose $\sum a_n$ is a series such that*

$$\lim_{n\to\infty} \frac{|a_{n+1}|}{|a_n|} = q.$$

Then $q \geq 0$, and

$$\sum a_n \text{ converges if } q < 1;$$
$$\sum a_n \text{ diverges if } q > 1.$$

Proof. If $q < 1$ then it follows from Proposition 9.2.3 that the series $\sum |a_n|$ converges, Hence, by Proposition 9.3.2, the given series $\sum a_n$ also converges. If $q > 1$ then for almost all n $|a_{n+1}| > |a_n|$ so that $\{a_n\}$ does not converge to 0. Consequently, by Proposition 9.1.3.3, $\sum a_n$ diverges. $\qquad\qquad\square$

Needless to say, the evaluation of the limit of a convergent series, an issue sidestepped in the foregoing discussion, is quite frequently of interest. When this happens, the said limit can always be estimated by its partial sums. However, to be useful, such a partial sum should be accompanied by some appraisal of the accuracy of this approximation.

Example 9.3.6 The difference between $\sum 1/n^n$ and the partial sum

$$1 + \frac{1}{2^2} + \frac{1}{3^3} + \frac{1}{4^4} + \frac{1}{5^5} + \frac{1}{6^6} + \frac{1}{7^7} + \frac{1}{8^8} + \frac{1}{9^9}$$

is

$$\frac{1}{10^{10}} + \frac{1}{11^{11}} + \frac{1}{12^{12}} + \cdots < \frac{1}{10^{10}} + \frac{1}{10^{11}} + \frac{1}{10^{12}} + \cdots$$

$$= \frac{1}{10^{10}}\left(1 + \frac{1}{10} + \frac{1}{10^2} + \cdots\right) = \frac{1}{10^{10}} \cdot \frac{10}{9}$$

$$= \frac{1}{9\cdot 10^9} = .000,000,000,111\ldots .$$

Hence, since the above partial sum is $1.291,285,996,959,\ldots$, it follows that

$$1.291,285,996 < \sum \frac{1}{n^n} < 1.291,285,998.$$

A similar argument (Exercise 10) leads to the conclusion that the difference between the infinite series $\sum 1/n^n$ and its partial sum $1 + 1/2^2 + \cdots + 1/m^m$ is less than $1/(m(m+1)^m)$.

Example 9.3.7 Show that the difference between the infinite series $\sum \frac{\sin nx}{n^3}$ and the partial sum of its first m terms is less than $1/(2m^2)$. We use an integral comparison argument that is nearly identical with that used in the proof of Proposition 9.1.2. The absolute value of the incurred error is

$$\left| \frac{\sin{(m+1)x}}{(m+1)^3} + \frac{\sin{(m+2)x}}{(m+2)^3} + \frac{\sin{(m+3)x}}{(m+3)^3} + \cdots \right|$$

$$\leq \left| \frac{\sin{(m+1)x}}{(m+1)^3} \right| + \left| \frac{\sin{(m+2)x}}{(m+2)^3} \right| + \left| \frac{\sin{(m+3)x}}{(m+3)^3} \right| + \cdots$$

$$\leq \frac{1}{(m+1)^3} + \frac{1}{(m+2)^3} + \frac{1}{(m+3)^3} + \cdots < \int_m^\infty \frac{dx}{x^3} = \frac{1}{2m^2}.$$

It should be stressed, however, that approximating a series by its partial sums can be inefficient to the point of rendering the partial sums useless. Such, for example, is the case with the infinite series solution of the important three body problem of physics. Another explicit instance is given in Example 9.3.9 below.

A series $\sum a_n$ for which $\sum |a_n|$ is convergent is said to be *absolutely convergent*. It is clear that for series with positive terms absolute convergence is equivalent to mere convergence. In general, however, that is not the case. The series

$$\sum \frac{(-1)^{n-1}}{n} = 1 - \frac{1}{2} + \frac{1}{3} - \frac{1}{4} + \cdots \tag{11}$$

is not absolutely convergent because $\sum |(-1)^{n-1}/n| = \sum 1/n$ which, by Proposition 9.1.4, is divergent. Nevertheless, it follows from the next proposition that the series (11) is convergent.

Proposition 9.3.8 *Suppose $\{a_n\}$ is a decreasing sequence that converges to 0. Then the series $\sum (-1)^n a_n$ converges.*

Proof. Define e_n and f_n to be the partial sums

$$e_n = a_0 - a_1 + a_2 - a_3 + \cdots + a_{2n-2} - a_{2n-1} + a_{2n}$$
$$f_n = a_0 - a_1 + a_2 - a_3 + \cdots - a_{2n-3} + a_{2n-2} - a_{2n-1}$$

Since $a_{2n-2} \geq a_{2n-1} \geq a_{2n} \geq 0$ it follows that

$$e_n = e_{n-1} - (a_{2n-1} - a_{2n}) \leq e_{n-1}$$
$$f_n = f_{n-1} + (a_{2n-2} - a_{2n-1}) \geq f_{n-1}$$
$$e_n = f_n + a_{2n} \geq f_n$$

and hence

$$e_1 \geq e_2 \geq \cdots \geq e_n \geq f_n \geq \cdots f_2 \geq f_1 \qquad \text{for } n = 1, 2, 3, \ldots.$$

Hence both of the sequences $\{e_n\}$ and $\{f_n\}$, being monotone and bounded, converge. Moreover $e_n - f_n = a_{2n}$ and $\{a_{2n}\} \to 0$, so that Proposition 8.1.4

guarantees that $\{e_n\}$ and $\{f_n\}$ converge to the same limit. Since $\{e_n\}$ and $\{f_n\}$ constitute all the partial sums of the series $\sum(-1)^n a_n$ it follows that this series converges.   □

Example 9.3.9 In Section 3.2 it was argued informally that

$$\frac{\pi}{4} = 1 - \frac{1}{3} + \frac{1}{5} - \frac{1}{7} + \frac{1}{9} - \frac{1}{11} + \cdots .$$

It can now be asserted on the basis of the above proposition that this alternating series does indeed converge. However, if d_n denotes the nth partial sum, then

$$d_1 = 1,$$
$$d_{10} = .76045\ldots ,$$
$$d_{100} = .78289\ldots ,$$
$$d_{1,000} = .78514\ldots ,$$
$$d_{10,000} = .78537\ldots ,$$

indicating an annoyingly slow convergence of the partial sums. A formal proof that this series converges to $\pi/4$ will be offered in Section 14.2. It is important to note that without the alternation of signs this series would diverge.

A series whose terms alternate in sign is said to be *alternating*. The alternating series of Exercise 5 indicates that the hypotheses of Proposition 9.3.8 are indeed essential.

When faced with the problem of determining whether a given series converges or diverges the reader may find the chart in Figure 9.3 helpful.

Exercises 9.3

1. Decide whether the following series converge or diverge. Justify your answer.

a) $\sum(-1)^n \dfrac{n}{n+2}$

b) $\sum(-1)^n \dfrac{1}{2n+3}$

c) $\sum(-1)^n \dfrac{1}{n+1}$

d) $\sum(-1)^n \dfrac{1}{\sqrt{n}}$

e) $\sum \dfrac{n^2}{n^2+1}$

f) $\sum \dfrac{(-1)^n n^2}{n^2+1}$

g) $\sum \dfrac{n^3}{n^2+1}$

h) $\sum \dfrac{(-1)^n n^3}{n^2+1}$

i) $\sum \dfrac{n}{n^2+1}$

j) $\sum \dfrac{(-1)^n n}{n^2+1}$

k) $\sum \cos n\pi$

l) $\sum(-1)^n \cos n\pi$

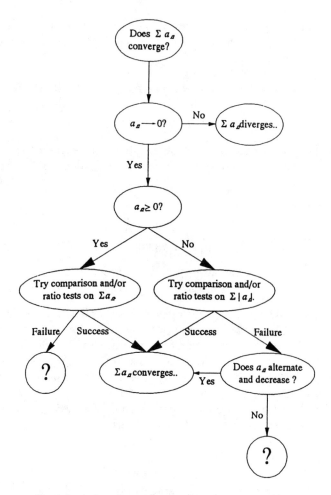

Fig. 9.3 A flowchart for the convergence of a series.

m) $\sum \dfrac{\cos n\pi}{n}$

n) $\sum \dfrac{(-1)^n \cos n\pi}{n}$

o) $\sum \dfrac{\cos n\pi}{n^2}$

p) $\sum \dfrac{(-1) \cos x}{n^2}$

q) $\sum \dfrac{\cos nx}{n^2}$

r) $\sum \dfrac{(-1)^n \cos nx}{n^2}$

2. For what values of p does $1 - 1/2^p + 1/3^p - 1/4^p + \cdots$ converge?

3. Prove that if $\sum a_n^2$ and $\sum b_n^2$ both converge, then $\sum a_n b_n$ also converges.

4. True or false: if $\sum a_n$ converges, then $\sum a_n^2$ also converges? Justify your answer.

5. Define $a_{2n} = 1/(n+1)$ and $a_{2n+1} = 1/2^n$ for $n = 0, 1, 2, \ldots$. Prove that $\sum(-1)^n a_n$ diverges.

6. Prove that the series

$$1 + \frac{1}{2} - \frac{1}{3} + \frac{1}{4} + \frac{1}{5} - \frac{1}{6} + \frac{1}{7} + \frac{1}{8} - \frac{1}{9} + \cdots$$

diverges.

7. Let $\sum a_n$ be a convergent series with $a_n \geq 0$, and let $\{b_n\}$ be a bounded sequence. Prove that $\sum a_n b_n$ is also convergent.

8. Prove that if the series $\sum a_n$ converges, then so does the series $\sum a_n/(1 + a_n)$.

9. Let $\sum a_n$ be a convergent series with $a_n \leq 0$, $\sum b_n$ a convergent series with $b_n \geq 0$, and $\sum c_n$ a series with $a_n \leq c_n \leq b_n$ for $n = 0, 1, 2, \ldots$. Prove that $\sum c_n$ is also convergent.

10. Show that the difference between the infinite series $\sum 1/n^n$ and the partial sum of its first m terms is less than $1/(m(m+1)^m)$.

11. Show that the difference between the infinite series $\sum \sin\, nx/n^p$ and the partial sum of its first m terms is less than $1/((p-1)m^{p-1})$.

12. Suppose $\sum a_n$ is a series such that $\sqrt[n]{|a_n|} \to q$. Prove that $\sum a_n$ converges (absolutely) if $q < 1$ and diverges if $q > 1$.

13. Prove that the series $\sum x(1-x)^n(nx + x - 1)$ converges for $x \in [0, 1]$.

14. Prove that the series

$$1 + \frac{1}{2^2} - \frac{1}{3^2} - \frac{1}{4^2} + \frac{1}{5^2} + \frac{1}{6^2} - \frac{1}{7^2} - \frac{1}{8^2} + \cdots$$

converges.

15. Prove that the series

$$1 + \frac{1}{2} - \frac{1}{3} - \frac{1}{4} + \frac{1}{5} + \frac{1}{6} - \frac{1}{7} - \frac{1}{8} + \cdots$$

converges.

16. Prove that the series

$$1 - \frac{1}{2} - \frac{1}{3} - \frac{1}{4} + \frac{1}{5} - \frac{1}{6} - \frac{1}{7} - \frac{1}{8} + \cdots$$

converges.

17. Suppose $\sum a_n$ and $\sum b_n$ are convergent series. If $c_n = \max\{a_n, b_n\}$, does the series $\sum c_n$ necessarily converge? Justify your answer.

18. Suppose $\sum a_n$ is a convergent series where $a_n \geq a_{n+1} \geq 0$ for all n. Prove that $\lim_{n \to \infty} n \cdot a_n = 0$.

19. Prove that

$$\sum_{n=1}^{m} \frac{x^{2^{n-1}}}{1 - x^{2^n}} = \frac{1}{1 - x} - \frac{1}{1 - x^{2^m}}.$$

Conclude that

$$\sum_{n=1}^{\infty} \frac{x^{2^{n-1}}}{1 - x^{2^n}}$$

converges for all $x \neq \pm 1$. Find the sum.

20. Prove that

$$\sum_{n=1}^{m} \frac{x}{(nx + 1)(nx - x + 1)} = \frac{mx}{mx + 1}.$$

Evaluate

$$\sum_{n=1}^{\infty} \frac{x}{(nx + 1)(nx - x + 1)}$$

for each value of x.

21. Suppose $\sum a_n$ and $\sum b_n$ are both divergent series. Does $\sum(a_n + b_n)$ necessarily diverge? Justify your answer.

9.4* THE MOST CELEBRATED PROBLEM

The main role that the series $\sum 1/n^p$ of Proposition 9.1.2 plays in this context is to provide us with a tool for verifying the convergence of some trigonometric series , as was done in Example 9.3.4. However, this series is of great interest in other areas of mathematics as well and we shall now digress to show how both its divergence for $p = 1$ and convergence for $p = 2$ yield interesting information about integers in general and prime numbers in particular. The interest in prime numbers goes back at least as far as Euclid (circa 300 B.C.) whose famed book *The Elements* includes a proof of the fact that there is an infinite number of prime numbers. An alternate, and ultimately more fruitful, proof was offered by Euler two thousand years later and this one is paraphrased below.

Proposition 9.4.1 *There is an infinite number of prime integers.*

Proof. Suppose, by way of contradiction, that there is only a finite number of primes and that these are $2, 3, 5, \ldots, p$ where p is the largest prime. It follows from Proposition 9.1.1 that

$$1 + \frac{1}{2} + \frac{1}{2^2} + \frac{1}{2^3} + \frac{1}{2^4} + \cdots = \frac{1}{1 - 1/2} = 2$$

$$1 + \frac{1}{3} + \frac{1}{3^2} + \frac{1}{3^3} + \frac{1}{3^4} + \cdots = \frac{1}{1 - 1/3} = \frac{3}{2}$$

$$1 + \frac{1}{5} + \frac{1}{5^2} + \frac{1}{5^3} + \frac{1}{5^4} + \cdots = \frac{1}{1 - 1/5} = \frac{5}{4}$$

$$\cdots$$

$$1 + \frac{1}{p} + \frac{1}{p^2} + \frac{1}{p^3} + \frac{1}{p^4} + \cdots = \frac{1}{1 - 1/p} = \frac{p}{p - 1}.$$

Consequently the product

$$\left(1 + \frac{1}{2} + \frac{1}{2^2} + \frac{1}{2^3} + \frac{1}{2^4} + \cdots\right)\left(1 + \frac{1}{3} + \frac{1}{3^2} + \frac{1}{3^3} + \frac{1}{3^4} + \cdots\right)$$

$$\left(1 + \frac{1}{5} + \frac{1}{5^2} + \frac{1}{5^3} + \frac{1}{5^4} + \cdots\right) \cdots \left(1 + \frac{1}{p} + \frac{1}{p^2} + \frac{1}{p^3} + \frac{1}{p^4} + \cdots\right)$$

$$\tag{12}$$

is finite (it actually equals $2 \cdot (3/2) \cdot (5/4) \cdots (p/(p-1))$, but that turns out to be immaterial). Note, however, that if the multiplication implicit in Expression (12) is actually carried out , i.e., if the parentheses are opened, an infinite sum is obtained, each of whose summands has the form

$$\frac{1}{2^{d_2}} \cdot \frac{1}{3^{d_3}} \cdot \frac{1}{5^{d_5}} \cdots \cdots \frac{1}{p^{d_p}} = \frac{1}{2^{d_2} \cdot 3^{d_3} \cdot 5^{d_5} \cdots \cdots p^{d_p}} \tag{13}$$

where $d_2, d_3, d_5, \ldots, d_p$ are arbitrary non-negative integers. Moreover each of these summands will occur exactly once. The denominator of the right-side of Equation (13) is a positive integer and since each integer greater than 1 can be factored into primes in a unique manner it follows that each positive integer appears as the denominator of a summand exactly once. Consequently

$$\left(1 + \frac{1}{2} + \frac{1}{2^2} + \frac{1}{2^3} + \frac{1}{2^4} + \cdots\right)\left(1 + \frac{1}{3} + \frac{1}{3^2} + \frac{1}{3^3} + \frac{1}{3^4} + \cdots\right)$$

$$\left(1 + \frac{1}{5} + \frac{1}{5^2} + \frac{1}{5^3} + \frac{1}{5^4} + \cdots\right) \cdots \left(1 + \frac{1}{p} + \frac{1}{p^2} + \frac{1}{p^3} + \frac{1}{p^4} + \cdots\right)$$

$$= 1 + \frac{1}{2} + \frac{1}{3} + \frac{1}{4} + \frac{1}{5} + \frac{1}{6} + \frac{1}{7} + \frac{1}{8} + \cdots. \tag{14}$$

Since the left-side of Equation (14) is finite and the right side is known by Proposition 9.1.2 to diverge, we have obtained the desired contradiction. It follows that the number of prime integers is infinite.   □

This proof should be taken with a grain of salt since the opening of the parentheses in Expression (12) has not been formally justified. Nevertheless, this operation is intuitively plausible and can be rigorously justified without serious difficulty (see Exercise 9.2.9). We now go on to derive an interesting consequence from the convergence of $\sum 1/n^2$ to $\pi^2/6$.

Proposition 9.4.2 *If two positive integers are selected at random and independently of each other, then the probability that they are relatively prime is* $6/\pi^2$.

Informal proof. Let p be a fixed prime integer. It is then reasonable to assume that the probability that a randomly chosen integer is divisible by p is $1/p$. Consequently, if a and b are two randomly (and independently) chosen integers then

$$\text{Probability } [p \text{ divides both } a \text{ and } b] = \frac{1}{p^2}$$

and so

$$\text{Probability } [p \text{ does not divide both } a \text{ and } b] = 1 - \frac{1}{p^2}.$$

Let $p_1, p_2, p_3, \ldots, p_k, \ldots$ be a listing of the prime integers, and let P_k denote the probability that p_k does not divide both a and b. Assuming that divisibility by distinct p_i and p_j are independent events, it follows that if P denotes the probability that a and b are relatively prime then $P = P_1 \cdot P_2 \cdot P_3 \cdots \cdot P_k \cdots$ and so

$$\frac{1}{P} = \frac{1}{P_1 \cdot P_2 \cdot P_3 \cdots \cdot P_k \cdots}$$

$$= \frac{1}{1 - \frac{1}{2^2}} \cdot \frac{1}{1 - \frac{1}{3^2}} \cdot \frac{1}{1 - \frac{1}{5^2}} \cdots$$

$$= \left(1 + \frac{1}{2^2} + + \frac{1}{(2^2)^2} + \frac{1}{(2^2)^3} + \frac{1}{(2^2)^4} + \cdots\right)$$

$$\cdot \left(1 + \frac{1}{3^2} + \frac{1}{(3^2)^2} + \frac{1}{(3^2)^3} + \frac{1}{(3^2)^4} + \cdots\right)$$

$$\cdot \left(1 + \frac{1}{5^2} + + \frac{1}{(5^2)^2} + \frac{1}{(5^2)^3} + \frac{1}{(5^2)^4} + \cdots\right) \cdots$$

$$= \left(1 + \frac{1}{2^2} + \frac{1}{(2^2)^2} + \frac{1}{(2^3)^2} + \frac{1}{(2^4)^2} + \cdots\right)$$

$$\cdot \left(1 + \frac{1}{3^2} + \frac{1}{(3^2)^2} + \frac{1}{(3^3)^2} + \frac{1}{(3^4)^2} + \cdots\right)$$

$$\cdot \left(1 + \frac{1}{5^2} + \frac{1}{(5^2)^2} + \frac{1}{(5^3)^2} + \frac{1}{(5^4)^2} + \cdots\right) \cdots$$

An argument analogous to that used in the proof of Proposition 9.4.1 now permits us to conclude that the above infinite product equals

$$1 + \frac{1}{2^2} + \frac{1}{3^2} + \frac{1}{4^2} + \frac{1}{5^2} + \frac{1}{6^2} + \cdots = \frac{\pi^2}{6}$$

Hence $P = 6/\pi^2$.   □

The foundations of the above proof are even shakier than those of Proposition 9.4.1. It assumes that it is possible to chose an integer at random, a subject that turns out to be fraught with logical perils. It also makes other unjustified, though plausible, probabilistic assumptions. Still, all these objections can be met at the cost of complicating the proof, and the conclusion is in fact valid.

Exercise 2 demonstrates another application of the series $\sum 1/n^2$. Exercises 3 and 4 below, in conjunction with results such as Exercises 5.1.4c, d, indicate that the limits of the series $\sum 1/n^k$ are interesting for larger k as well.

In his 1859 paper on prime numbers B. Riemann defined a differentiable complex function $\zeta(z)$ such that

$$\zeta(z) = 1 + \frac{1}{2^z} + \frac{1}{3^z} + \frac{1}{4^z} + \cdots$$

whenever the series on the right converges. This function is now known as the *zeta function*. Note that it follows from Proposition 5.2 that $\zeta(2) = \pi^2/6$, and from Proposition 9.1.2 that $\zeta(1)$ does not exist. Riemann demonstrated that the zeta function contains much information about the distribution of prime numbers. This claim is of course substantiated to some extent by both Propositions 9.4.1 and 9.4.2. Of particular interest, claimed Riemann, are the zeroes of the zeta function, in other words, those complex numbers s such that

$$\zeta(s) = 0.$$

He proved that all such complex numbers s must fall inside the infinite strip bounded by the vertical straight lines $x = 0$ and $x = 1$ (Fig. 9.4). This information was used by him to estimate the number of prime numbers that fall in any interval. The same technique was used by Jacques Hadamard (1865–1963) and Charles-Jean de la Vallée Poussin (1866–1962) to prove, independently of each other, the Prime Number Theorem in 1896. This theorem had already been conjectured by Carl Friedrich Gauss (1777–1855) and others over one hundred years earlier.

Theorem 9.4.3 (The Prime Number Theorem) *For every real number* x *let* $\pi(x)$ *denote the number of prime integers less than* x *and let* $Li(x) = \int_2^x dt/\ln t$. *Then*

$$\lim_{x \to \infty} \frac{\pi(x)}{Li(x)} = 1.$$

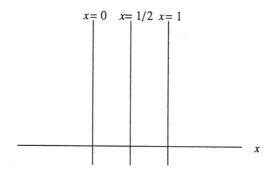

Fig. 9.4   The location of the zeros of the zeta function.

In effect, this theorem says that the number of primes not exceeding x is approximated by $\int_2^x dt/\ln t$. The table below, which comes from a letter written by Gauss, indicates that this approximation is fairly good.

x	$\pi(x)$	$Li(x)$	Difference
500,000	41,556	41,606.4	50.4
1,000,000	78,501	78,627.5	126.5
1,500,000	114,112	114,263.1	151.1
2,000,000	148,883	149,054.8	171.8
2,500,000	183,016	183,245.0	229.0
3,000,000	216,745	216,970.6	225.6

Riemann actually believed, and tried unsuccessfully to prove, that the strip containing all the zeroes of the zeta function could be narrowed down to the straight line $x = 1/2$ which bisects it. This conjecture, better known as *Riemann's Zeta Hypothesis*, is now considered by the mathematical establishment to be the most celebrated problem of pure mathematics.

Conjecture 9.4.4 (The Riemann Zeta Hypothesis) *If s is a complex number such that $\zeta(s) = 0$, then the real part of s is $1/2$.*

It is now known that the line $x = 1/2$ contains an infinite number of the zeroes of $\zeta(s)$ and that the first 70,000,000 such zeroes do indeed lie on that line. This conjecture, if true, would result in a sharpened estimate of the difference between $\pi(x)$ and $Li(x)$. More specifically, the validity of the Riemann Zeta Hypothesis would imply that for some constant K and for all sufficiently large x

$$\frac{|\pi(x) - Li(x)|}{\pi(x)} \leq \frac{K(\ln x)^2}{\sqrt{x}}.$$

Exercises 9.4

1. The Prime Number Theorem is frequently stated in an alternate form that asserts that $\pi(x)$ is approximated by $x/(\ln x)$. Use L'Hospital's rule to prove that

$$\lim_{x \to \infty} \frac{Li(x)}{x/(\ln x)} = 1.$$

2. An integer that is not divisible by the square of any prime number is said to be *square free*. Use the methods of this section to argue that the probability that a randomly selected integer is square free is $6/\pi^2$.

3. Use the methods of this section to argue that if k is a fixed positive integer and m is a randomly selected positive integer, then the probability that m is not divisible by p^k for every prime p is $\left(\sum n^{-k}\right)^{-1}$.

4. A set of integers is said to be relatively prime if the only positive integer that divides all of them is 1. Use the methods of this section to argue that if a set of k integers is selected at random then the probability that they are relatively prime is $\left(\sum n^{-k}\right)^{-1}$.

5. If a is a positive real number and $z = x + iy$ is a complex number, we define $a^z = a^x \left(\cos(y \ln a) + i \sin(y \ln a)\right)$. Use De Moivre's formula to prove that if z and w are two complex numbers then $a^z a^w = a^{z+w}$.

6. Describe the main mathematical achievements of
   a) Carl Friedrich Gauss
   b) Charles-Jean de la Vallée Poussin
   c) Jacques Hadamard

Chapter Summary

In this chapter, the results of Chapter 8 are applied to infinite series, and the convergence and divergence of some standard series are demonstrated. The nature of other series can be determined by comparing them with the standard ones. Special attention is given to series with positive terms and series with terms whose signs alternate. The last section informally relates the convergence of some series to the distribution of prime numbers.

10

Series of Functions

By and large the series examined in Chapter 9 had constant terms. This chapter examines series whose terms are functions. The discussion of such series will be limited to the question of convergence of Newton's Power Series and Euler's Trigonometric Series.

10.1 POWER SERIES

If $\{c_n\}$ is any sequence, then we define the series $\sum c_n x^n$ as a *power series*. For example, if $c_n = n^2 + 1$ for $n = 0, 1, 2, \ldots$ then

$$\sum c_n x^n = \sum \left(n^2 + 1\right) x^n = 1 + 2x + 5x^2 + 10x^3 + 17x^4 + \ldots .$$

For a given sequence $\{c_n\}$, the power series $\sum c_n x^n$ is a function (of the variable x) whose domain consists of those values of x for which the series converges. With respect to the issue of the convergence of power series, the behavior of the geometric series $\sum x^n$ is fairly typical. Propositions 9.1.1 and 9.1.5 state that the series $\sum x^n$ converges if and only if $|x| < 1$. Thus, the domain of the function $\sum x^n$ consists of the open interval $(-1, 1)$. We now show that all power series exhibit a similar behavior.

Lemma 10.1.1 *Suppose the power series $\sum c_n x^n$ converges for $x = x'$ and diverges for $x = x''$, then $\sum c_n x^n$*

1. *converges absolutely for each x such that $|x| < |x'|$;*

2. *diverges for each x such that $|x| > |x''|$.*

Proof of 1. If $x' = 0$ then there is nothing to prove, and so we may assume that $x' \neq 0$. Since $\sum c_n x'^n$ converges it follows that $\{c_n x'^n\}$ converges (to 0) and so $\{c_n x'^n\}$ is bounded. In other words, there exists a positive number M such that

$$|c_n x'^n| \leq M \qquad \text{for } n = 0, 1, 2, \ldots .$$

Let x be any number such that $|x| < |x'|$. Then

$$|c_n x^n| = |c_n x'^n| \cdot \left|\frac{x}{x'}\right|^n \leq M \left|\frac{x}{x'}\right|^n .$$

Since $|x/x'| < 1$, the geometric series $\sum M|x/x'|^n$ converges. Hence, by Proposition 9.2.1, the series $\sum |c_n x|^n$ converges and by Proposition 9.3.2 $\sum c_n x^n$ also converges.

Proof of 2: This is proved by contradiction. Suppose x is such that $|x| > |x''|$ and $\sum c_n x^n$ converges. It follows from part 1 of this proof that $\sum c_n x''^n$ also converges, contrary to assumption. Hence, $\sum c_n x^n$ diverges. □

It should be noted that the strict inequalities in conclusions 1 and 2 of Lemma 10.1.1 are unavoidable. To see this, examine the series $\sum x^n/n$ which converges for $x' = -1$ and diverges for $x'' = 1$. Thus, even though $|x'| = |x''| = 1$, the convergence of the series at x' cannot be used to prove its convergence at x'' and, vice versa, the divergence of the series at x'' cannot be used to prove its divergence at x'.

Theorem 10.1.2 *Let $\sum c_n x^n$ be a power series. Then one of the following must hold:*

1. *$\sum c_n x^n$ converges absolutely for every value of x;*

2. *$\sum c_n x^n$ converges only for $x = 0$*

3. *there is a number $\rho > 0$ such that $\sum c_n x^n$ converges absolutely whenever $|x| < \rho$ and diverges whenever $|x| > \rho$.*

Proof. Let S be the set of all x's for which the given series converges. Clearly $0 \in S$. If the set S is unbounded, then it follows from Lemma 10.1.1.1 that the series converges absolutely for every x, so that alternative 1 holds. Hence it may be assumed that S is bounded. It follows from the Completeness Axiom that S has a least upper bound, say ρ.

If $x > \rho$, then $x \notin S$ and so the series diverges at x.

If $x < -\rho$ let x_0 be such that $x < x_0 < -\rho$ (see Fig. 10.1). Then $\rho < -x_0 < -x$. It follows from the definition of ρ that the series diverges at $-x_0$ and so, by Lemma 10.1.1.2, the series also diverges at x because $|x| = -x > -x_0$.

Fig. 10.1   A proof of divergence.

Finally, suppose, by way of contradiction, that the series diverges for some x'' such that $|x''| < \rho$. It then follows from Lemma 10.1.1 that the series diverges for all $x > |x''|$, i.e, $|x''|$ is an upper bound of S. This contradicts the fact that ρ is the <u>least</u> upper bound of S and hence the series converges for all x such that $|x| < \rho$. That this convergence is absolute follows from the fact that for any x such that $|x| < \rho$, there clearly exists an x' such that $|x| < x' < \rho$. For, by Lemma 10.1.1, convergence at x' entails absolute convergence x.   □

If alternative 3 holds, ρ is called the *radius of convergence* of the given power series (this terminology is borrowed from the calculus of complex variables where the set of those values of the independent variable for which the power series converges consists of a disk). In this case, the set of all x's for which the power series converges is an interval of the form $(-\rho, \rho)$, $[-\rho, \rho)$, $(-\rho, \rho]$, or $[-\rho, \rho]$ and all the possiblities can occur (see Exercises 10–13). This is called the *interval of convergence*. If alternatives 1 or 2 hold, the series is said to have *infinite* $(\rho = \infty)$ or *zero* $(\rho = 0)$ *radius of convergence*, respectively.

It is interesting that the proof of Lemma 10.1.1 is based on a comparison of the power series with the infinite geometric progression. To some extent this vindicates Newton's insufficient discussion of this issue in the concluding paragraphs of *De Analysi* (Section 3.3).

The proof of Theorem 10.1.2 is existential in nature and provides no method for deriving the radius of convergence. The next proposition shows how this radius can be computed for many power series.

Proposition 10.1.3 *Suppose $\sum c_n x^n$ is a power series with radius of convergence ρ. If $c_n \neq 0$ for all n and*

$$\left\{ \frac{c_{n+1}}{c_n} \right\} \to q$$

then

1. $\rho = \infty$ *if $q = 0$,*

2. $\rho = 1/|q|$ *if $q \neq 0$.*

Proof. Since $\{c_{n+1}/c_n\} \to q$ it follows that for $x \neq 0$

$$\left\{ \frac{|c_{n+1} x^{n+1}|}{|c_n x^n|} \right\} \to |qx|.$$

Hence, by Proposition 9.2.3, the infinite series $\sum |c_n x^n|$

$$\text{converges for all } x \text{ such that } |qx| < 1 \tag{1}$$
$$\text{diverges for all } x \text{ such that } |qx| > 1 \tag{2}$$

If $q = 0$ then it follows from (1) that $\rho = \infty$. On the other hand, if $q \neq 0$, then it follows from (1), (2) that the series converges or diverges according as $|x| < 1/|q|$ or $|x| > 1/|q|$. Hence $\rho = 1/|q|$. □

Example 10.1.4 For the series $\sum x^n/(n2^n)$ we have

$$\frac{c_{n+1}}{c_n} = \frac{n2^n}{(n+1)2^{n+1}} = \frac{n}{2(n+1)} \to \frac{1}{2}.$$

Consequently, this series has radius of convergence 2. For $x = 2$ this series becomes the divergent series $\sum 1/n$ (Proposition 9.1.2) and for $x = -2$ this series becomes the convergent series $-1 + 1/2 - 1/3 + 1/4\ldots$ (Proposition 9.3.8).

Example 10.1.5 Newton's as yet unproved Fractional Binomial Theorem 3.1.1 states that for every rational number r

$$(1 + x)^r = \sum \binom{r}{k} x^k \tag{3}$$

where $\binom{r}{0} = 1$ and

$$\binom{r}{k+1} = \binom{r}{k} \cdot \frac{r-k}{k+1} \qquad \text{for all } k = 0, 1, 2, \ldots . \tag{4}$$

At this point it is possible to compute the radius of convergence ρ of the binomial series. It is clear that if r is not a natural number then

$$\frac{r-k}{k+1} \neq 0 \qquad \text{for all } k = 0, 1, 2, \ldots$$

It follows from Equation (4) by a straightforward induction that $\binom{r}{k} \neq 0$ for all k and hence Proposition 10.1.3 is applicable. Since

$$\lim_{k \to \infty} \frac{\binom{r}{k+1}}{\binom{r}{k}} = \lim_{k \to \infty} \frac{r-k}{k+1} = -1$$

it follows that if r is not a natural number then the binomial series has radius of convergence $\rho = 1$.

If r is a natural number then $\binom{r}{r+1} = \binom{r}{r} \cdot 0 = 0$ so that the binomial series contains only $r + 1$ non-zero terms. In this case the radius of convergence is clearly infinite.

The behavior of a power series at the endpoints $\pm\rho$ of its interval of convergence varies from series to series (Exercises 10–13). It is also easily verified that the series $\sum n^n x^n$ converges only for $x = 0$ (Exercise 8) whereas the series $\sum x^n/n^n$ converges for every value of x (Exercise 9).

Exercises 10.1

1. For which values of x does the power series $\sum_{n=1}^{\infty} c_n x^n$ converge, where the value of c_n is given below? Justify your answers.

a) $\dfrac{1}{n}$

b) n

c) $\dfrac{1}{\sqrt{n}}$

d) $\sqrt{n}$

e) n^2

f) $\dfrac{1}{n^2}$

g) $\dfrac{n}{n!}$

h) $\dfrac{n^{10}}{n!}$

i) $\dfrac{1}{n+17}$

j) $\dfrac{(-1)^n}{an+b}$

k) $\dfrac{(-1)^n}{\ln(n+1)}$

l) $\dfrac{(-1)^n}{\ln(\ln(10n))}$

m) $\sqrt[4]{n}$

n) $\dfrac{1}{\sqrt[4]{n}}$

o) a^n

p) a^{n^2}

q) $5n$

r) 5^n

s) n^5

t) $\dfrac{5}{n}$

u) $\dfrac{n}{5}$

v) $(n!)^5$

w) $n^5!$

2. Show that the series $\sum x^n/(2n+5)$ has interval of convergence $[-1, 1)$.

3. Give an example of a power series with interval of convergence $(-1, 1]$.

4. Show that the series $\sum x^n$ diverges at both endpoints of its interval of convergence.

5. Show that the series $\sum x^n/n^2$ converges at both endpoints of its interval of convergence.

6. Does there exist a power series which converges at $x = -1$ and diverges at $x = 2$? Justify your answer.

7. Does there exist a power series which diverges at $x = -1$ and converges at $x = 2$? Justify your answer.

8. Prove that the power series $\sum n^n x^n$ converges only for $x = 0$.

9. Prove that the power series $\sum x^n/n^n$ converge for all values of x

In Exercises 10–13, a is a positive real number.

10. Give an example of a power series with interval of convergence $[-a, a)$.

11. Give an example of a power series with interval of convergence $(-a, a]$.

12. Give an example of a power series with interval of convergence $(-a, a)$.

13. Give an example of a power series with interval of convergence $[-a, a]$.

14. Suppose r, s, a_0, a_1 are any real numbers and the sequence $\{a_n\}$ is defined by the recurrence $a_{n+2} = ra_{n+1} + sa_n$ for all n.

a) Prove that the power series $f(x) = \sum a_n x^n$ has a positive radius of convergence ρ.

b) Prove that there exist polynomials $P(x)$, $Q(x)$, of degree at most 2, such that $f(x) = P(x)/Q(x)$ for $|x| < \rho$.

c) Suppose $Q(x)$ has distinct zeros. Prove that there exist real numbers a, b, c, d such that $a_n = a \cdot b^n + c \cdot d^n$ for all n.

d) Find an explicit expression for a_n in the case where $Q(x)$ has only one repeated zero.

e) The Fibonacci numbers are defined by $F_0 = 0$, $F_1 = 1$, $F_{n+2} = F_{n+1} + F_n$ for all n. Find the numbers a, b, c, d whose existence is mentioned in part c above.

15. Let $\sum c_n x^n$, $\sum a_n x^n$ be power series with radii of convergence ρ_c, ρ_a respectively. Prove that if $|c_n| \geq |a_n|$ for almost all n, then $\rho_a \geq \rho_c$. Is it necessary that $\sum a_n x^n$ converge everywhere that $\sum c_n x^n$ does? Justify your answer.

16. Let $\sum c_n x^n$ be a series such that $\lim_{n \to \infty} \sqrt[n]{|c_n|} = L$. Prove that $\rho = 1/L$.

17. Prove that the Hypergeometric Series of Exercise 9.2.5 converges for $-1 < x < 0$ and diverges for $x < -1$.

18. Suppose there exist two polynomials $P(x)$, $Q(x)$ such that $c_n = P(n)/Q(n)$. Prove that the series $\sum c_n x^n$ has $\rho = 1$.

19. Suppose c_n is the digit in the nth place of the decimal expansion of $1/7$. What is the radius of convergence of $\sum c_n x^n$? For which values of x can you assert that the series converges?

20. Suppose c_n is the digit in the nth place of the decimal expansion of $\sqrt{2}$. What is the radius of convergence of $\sum c_n x^n$? For which values of x can you assert that the series converges?

21. Suppose c_n is the digit in the nth place of the decimal expansion of $1/7$. What is the radius of convergence of $\sum c_n^n x^n$? For which values of x can you assert that the series converges?

22. Suppose c_n is the digit in the nth place of the decimal expansion of $\sqrt{2}$. What is the radius of convergence of $\sum c_n^n x^n$? For which values of x can you assert that the series converges?

10.2 TRIGONOMETRIC SERIES

Just like power series, the domain of convergence of trigonometric series depends on the size of the coefficient c_n for large n. In order to quantify this notion, given two sequences $\{a_n\}$ and $\{b_n\}$, we say that $a_n = O(b_n)$ provided that $\lim_{n\to\infty} a_n/b_n$ exists. Thus,

$$\sqrt{n^2+1} = O(n) \qquad \text{because} \qquad \lim_{n\to\infty} \frac{\sqrt{n^2+1}}{n} = 1$$

$$\frac{n+1}{3n^3 - 5n^2 + 17} = O(n^{-2})$$

$$\text{because } \lim_{n\to\infty} \frac{\dfrac{n+1}{3n^3 - 5n^2 + 17}}{n^{-2}} = \lim_{n\to\infty} \frac{n^2(n+1)}{3n^3 - 5n^2 + 17} = \frac{1}{3}$$

Theorem 10.2.1 *Let $\{c_n\}$ be a sequence such that $c_n = O(n^q)$ for some $q < -1$. Then the series $\sum c_n \cos nx$ and $\sum c_n \sin nx$ both converge (absolutely) for each value of x.*

Proof. Since $\{c_n/n^q\}$ converges, it must be bounded , say

$$|c_n/n^q| \le M \qquad \text{for all } n.$$

Hence,

$$|c_n \cos nx| \le |c_n| \le Mn^q. \tag{2}$$

and so it follows from Proposition 9.1.2 that $\sum c_n \cos nx$ converges absolutely for each value of x. It is clear that the same proof also works for $\sum c_n \sin nx$. $\qquad \square$

Example 10.2.2 The series $\sum (\sqrt{n}/(n^2+1)) \cos nx$ converges for every value of x because $\sqrt{n}/(n^2+1) = O(n^{-1.5})$.

When the hypothesis of Theorem 10.2.1 is not satisfied the convergence of the series will in general depend on the value of x. Thus, the series $\sum \cos nx/n$ diverges for $x = \pi$ and converges for $x = 2\pi$ since

$$\sum \frac{\cos n\pi}{n} = -1 + \frac{1}{2} - \frac{1}{3} + \frac{1}{4} - \cdots \tag{5}$$

and

$$\sum \frac{\cos n(2\pi)}{n} = 1 + \frac{1}{2} + \frac{1}{3} + \frac{1}{4} + \cdots . \tag{6}$$

In fact $\sum(\cos nx)/n$ has the same convergent expansion of Equation (5) whenever x is an *odd* multiple of π, and the same divergent expansion of Equation (6) whenever x is an *even* multiple of π. Thus, for a particular trigonometric series, the values of x for which it converges may interlace with those values for which it diverges. This, of course, stands in marked contrast with the state of affairs for power series. It is clear from Theorem 10.1.2 that if a power series converges for x_1 and x_2 then it also converges for every x between x_1 and x_2. Hence, no such interlacing can occur for power series.

It was demonstrated in Example 9.1.4 that the series $\sum \cos nx$ diverges whenever x is a rational multiple of π. It can also be shown (Exercise 9.1.3) that this series diverges for every value of x. Once again, this stands in marked contrast with power series which must converge at least for $x = 0$. In general the behavior of trigonometric series is very complicated. It has been the subject of much research and has helped bring into being several branches of mathematics.

The following proposition will be helpful in deciding on the convergence of many particular trigonometric series. It's proof follows in a straightforward manner from Propositions 7.2.12 and 7.2.15.

Proposition 10.2.3 *Suppose r is a rational number, $a_0 > 0$, and $P(n) = a_0 n^d + a_1 n^{d-1} + \cdots + a_d$. Then $\{[P(n)]^r\} = O(n^{rd})$.*

Proof. See Exercises 3, 4.

Exercises 10.2

1. For which of the following sequences $\{c_n\}$ does the series $\sum c_n \cos nx$ converge for all x? Justify your answer.

a) n^2

b) $\sqrt{n+1}$

c) $(-1)^n/n^2$

d) $1/\sqrt{n+1}$

e) $\sqrt[3]{n^5+1}$

f) $(-1)^n/\sqrt[3]{n^5+1}$

g) $\sqrt[7]{n^2+1}$

h) $1/\sqrt[7]{n^2+1}$

i) $(-1)^n(n+n^{-1})^{-1}$

2. For which values of α can you guarantee that the series $\sum_{n=1}^{\infty} c_n \sin nx$ converges for all x, where c_n is given below? Justify your answers.

a) $n^{-\alpha}$

b) $\dfrac{1}{(n+2)^\alpha}$

c) $\dfrac{n^3+1}{n^\alpha+n}$

d) $\dfrac{n^\alpha+1}{n^{\alpha+1}}$

e) $\dfrac{n^\alpha+2}{n^{\alpha+2}}$

f) α^n

g) $\dfrac{\alpha^n}{n!}$

h) $\dfrac{\alpha^n+n^\alpha}{n^n}$

i) $\dfrac{\alpha^n+n^\alpha}{n!}$

3. Prove Proposition 10.2.3 for $r = 1$.

4. Prove Proposition 10.2.3.

In the exercises below you may assume that a_n, b_n, c_n, d_n are all positive numbers.

5. Prove that $a_n = O(a_n)$.

6. Prove that if $a_n = O(b_n)$ and $b_n = O(c_n)$, then $a_n = O(c_n)$.

7. Find $\{a_n\}$ and $\{b_n\}$ such that $a_n = O(b_n)$ but $b_n \neq O(a_n)$.

8. Prove that if $a_n = O(b_n)$ and $c_n = O(d_n)$ then $a_n c_n = O(b_n d_n)$.

9. Find $\{a_n\}$ and $\{b_n\}$ such that $a_n = O(b_n)$ but $1/a_n \neq O(1/b_n)$

Chapter Summary

The information garnered in Chapter 9 about infinite series is now applied to the question of determining the domain within which a given infinite series of functions converges. It is shown that in the case of Newton's power series this domain consists of a possibly infinite interval. The behavior of Euler's trigonometric series is much more complicated and so it is only demonstrated that for a large class of such series this domain extends to the whole real line.

11

Continuity

Before going on to differentiation it is necessary to examine closely the related notion of continuity.

11.1   AN INFORMAL INTRODUCTION

At first it was taken for granted that every function was necessarily continuous and differentiable, and during the two centuries following the birth of calculus it was quite common for mathematicians to use the terms continuity and differentiability more or less interchangeably. This confusion was accompanied and facilitated by imprecise definitions. Euler's definition,

> A *continuous* curve is one such that its nature can be expressed by a single function of x,

is a case in point. A *discontinuous* curve was one that had to be defined by different functions at different places. Euler was in all likelihood aware of the insufficiency of these definitions. In 1787, shortly after his death, the St. Petersburg Academy (where Euler had spent many of his working years) proposed the question of just how discontinuous the solutions to differential equations could be as a prize problem. A more concrete impetus for the clarification of continuity in analysis was provided by the problem of the vibrating string. As was mentioned in Chapter 5, it was in this context that mathematicians were forced for the first time to give serious consideration to functions whose graphs had corners that could not be ignored. The important role that discontinuous functions played in Fourier's 1807 award winning essay on heat must also have added to the evolving interest in the foundations of calculus.

The first satisfactory definition of continuity appeared in Bolzano's 1817 essay. His work did not receive the attention it deserved and the credit for actually turning the course of mathematics in the direction of more rigor goes to Cauchy's *Cours d'Analyse* of 1821. To Cauchy, continuity seems to have been a prerequisite for differentiability. After all, if the graph of the function $f(x)$ is to possess a tangent at a certain point, then the graph must consist of a single continuous arc in the vicinity of that point. More precisely, if the quotient

$$\frac{f(x+h) - f(x)}{h}$$

whose limit is the derivative $f'(x)$, is to converge as h approaches 0, then it is necessary that the numerator approach 0. Otherwise, the vanishing of the denominator would cause the fraction to diverge to infinity.

The notion of the limit of a sequence, a natural outgrowth of the issue of the convergence of infinite series, turned out to be extremely valuable in providing a language that made possible a careful distinction between the concepts of continuity and differentiability. Limits of sequences are used in the sequel to define first limits of functions and then continuity. It is also demonstrated that the property of continuity has several important implications regarding solutions of equations and the existence of maxima and minima.

11.2   THE LIMIT OF A FUNCTION

In this section and the next we present Cauchy's sequential definition of continuity. Loosely speaking, a function is continuous if its value at any point a is consistent with its values near a. We begin by formalizing the notion of "values near a". Let $f : D \to \mathbb{R}$ be a real valued function. If $a \in D$, we say that $\lim_{x \to a} f(x) = A$ provided that

Condition 1:
there is a sequence $\{x_n\} \subset D - \{a\}$ such that $\{x_n\} \to a$
and
Condition 2:
for every sequence $\{x_n\} \subset D - \{a\}$ such that $\{x_n\} \to a$, $\{f(x_n)\} \to A$.

Within the conventions of this text Condition 1 is tantamount to saying that the domain D contains some open interval of the form (a, b) or (b, a) (or both). Thus, for the function $f(x) = (4 - x^2)^{-1/2}$ with domain $(-2, 2)$, a could be any number in the closed interval $[-2, 2]$. In terms of the aforementioned informal definition of continuity, this condition guarantees that f does indeed have values near a. Condition 2 says that near a the function f assumes values close to A. This condition contains the substance of the notion of a limit.

Condition 1, on the other hand, will in general not be discussed explicitly — it is too easily verified.

Let r be a fixed real number and suppose f is the *constant function*

$$f(x) = r \quad \text{for all} \quad x \in \mathbb{R}.$$

Note that for every sequence $\{x_n\}$ whatsoever, convergent or not,

$$\{f(x_n)\} = \{r\} \to r.$$

Hence Condition 2 is automatically satisfied and so

$$\lim_{x \to a} f(x) = r$$

for every real number a.

Let f be the *identity function*

$$f(x) = x \quad \text{for all} \quad x \in R.$$

Then, if $\{x_n\} \to a$ we also have

$$\{f(x_n)\} = \{x_n\} \to a.$$

Again, Condition 2 is satisfied and so

$$\lim_{x \to a} f(x) = a.$$

These are useful observations and we state them together with another one as a proposition.

Proposition 11.2.1

1. $\lim\limits_{x \to a} r = r$

2. $\lim\limits_{x \to a} x = a$

3. $\lim_{x \to a} |x| = |a|$.

Proof. Limits 1 and 2 were proven above. The similarly easy proof of limit 3 is relegated to Exercise 12.   □

The following theorem, which is an analog of Theorem 7.2.3, now provides us with the means for calculating the limits of many functions.

Theorem 11.2.2 *Suppose $f, g : D \to \mathbb{R}$ are two functions such that*

$$\lim_{x \to a} f(x) = A \quad and \quad \lim_{x \to a} g(x) = B.$$

Then

1. $\lim\limits_{x \to a} (f(x) + g(x)) = A + B$

2. $\lim\limits_{x \to a} (f(x)g(x)) = A \cdot B$

3. $\lim\limits_{x \to a} (f(x) - g(x)) = A - B$

4. $\lim\limits_{x \to a} \dfrac{f(x)}{g(x)} = \dfrac{A}{B}$ provided $B \neq 0.$

Proof of 1. Suppose $\{x_n\} \subset D - \{a\}$ and $\{x_n\} \to a$. Then, by part 2 of Theorem 7.2.3, $\lim_{n \to \infty} (f(x_n) + g(x_n)) = A + B$ and hence

$$\lim_{x \to a} (f(x) + g(x)) = A + B.$$

Proof of 2. See Exercise 9.

Proof of 3 See Exercise 10.

Proof of 4. See Exercise 11.   □

It follows from the several parts of the·above theorem that

$$\lim_{x \to 2} \frac{x^3 - 3x^2 - x + 2}{x^2 - x} = \frac{2^3 - 3 \cdot 2^2 - 2 + 2}{2^2 - 2} = -2.$$

Note that we are not disturbed by the fact that the denominator function $x^2 - x$ happens to vanish for $x = 0, 1$. If and when necessary, we can restrict attention to any domain D that, while still containing 2, excludes 0 and 1. Such a domain could be the open interval $(3/2, 4)$.

By way of closure, it is fitting that the issue of $\lim_{x \to a} f(x)/g(x)$ when $g(x)$ vanishes at $x = a$, i.e., when $g(a) = 0$, be addressed. In such cases, the limit in question may or may not exist.

Example 11.2.3 Consider the function $f(x) = (x^2 - 9)/(x - 3)$ and the limit

$$\lim_{x \to 3} \frac{x^2 - 9}{x - 3}.$$

In this case the domain D does not include the number 3 since

$$f(3) = \frac{3^2 - 3}{3 - 3} = \frac{0}{0}$$

is undefined. Nevertheless, it is clear that the domain D of this function does contain sequences that converge to 3, for example, $\{3 + 1/n\}$. Now, if $\{x_n\} \subset D$ is any sequence converging to 3, then $x_n \neq 3$ for all n so that

$$\frac{x_n^2 - 9}{x_n - 3} = x_n + 3.$$

Hence

$$\lim_{n \to \infty} \frac{x_n^2 - 9}{x_n - 3} = \lim_{n \to \infty} (x_n + 3) = 3 + 3 = 6$$

and so

$$\lim_{x \to 3} \frac{x^2 - 9}{x - 3} = 6.$$

The readers may feel that the above example is concerned with a limit that is too artificial to merit consideration. Such is not the case. This is exactly the sort of limit whose resolution is required for the rigorous definition of the derivative that will be formulated in the next chapter.

Example 11.2.4 The expression $\lim_{x \to 3} 1/(x - 3)$ provides another instance of a limit of a rational function whose dominator vanishes at the crucial point. Here, the conclusion to be drawn is different from above. If D denotes the domain of this function, then $\{3 + (1/n)\} \to 3$, $\{3 + (1/n)\} \subset D - \{3\}$ and yet

$$\lim_{n \to \infty} f\left(3 + \frac{1}{n}\right) = \lim_{n \to \infty} \frac{1}{3 + \frac{1}{n} - 3} = \lim_{n \to \infty} n$$

which does not exist. Consequently, Condition 2 for the existence of limits does not hold, and so we say that $\lim_{x \to 3} 1/(x - 3)$ does not exist.

Limits can fail to exist for other reasons as well. Thus, $\lim_{x \to 3} \sqrt{1 - x^2}$ does not exist since the domain $D = [-1, 1]$ fails to have a sequence that converges to 3. The following examples are less trivial.

Example 11.2.5 The function $f(x) = |x|/x$, whose domain D consists of all the non-zero reals, fails to have a limit at 0. To see this note that

$$f(x) = \begin{cases} 1 & \text{if } x > 0 \\ -1 & \text{if } x < 0. \end{cases}$$

Consider the two sequences $\{1/n\}$ and $\{-1/n\}$ both of which converge to 0. It is clear that

$$\left\{ f\left(\frac{1}{n}\right) \right\} = \{1\} \to 1$$

and

$$\left\{ f\left(-\frac{1}{n}\right) \right\} = \{-1\} \to -1.$$

Since $1 \neq -1$ it follows that Condition 2 which requires that all the sequences in question have the <u>same</u> limit fails to hold for $f(x)$. Hence, here too, $\lim_{x \to 0} f(x)$ fails to exist.

Example 11.2.6 The *Dirichlet function* $f : \mathbb{R} \to \mathbb{R}$ is defined as:

$$f(x) = \begin{cases} 1 & \text{if } x \text{ is rational} \\ 0 & \text{if } x \text{ is irrational.} \end{cases} \tag{1}$$

This function has the property that $\lim_{x \to a} f(x)$ fails to exist at any a whatsoever. The reason for this lies in Proposition 7.2.16 according to which a is the limit of both a sequence $\{x_n\}$ of rationals and a sequence $\{\bar{x}_n\}$ of irrationals. Since

$$\{f(x_n)\} = \{1\} \to 1 \quad \text{and} \quad \{f(\bar{x}_n)\} = \{0\} \to 0$$

the inequality of these two limits implies the non-existence of $\lim_{x \to a} f(x)$.

Exercises 11.2

1. Decide whether the following limits exist and evaluate them when they do. Be sure to justify all of your steps.

a) $\displaystyle\lim_{x \to 5} \frac{x^2 - 25}{x - 5}$ b) $\displaystyle\lim_{x \to 5} \frac{x^2 - 25}{x + 5}$ c) $\displaystyle\lim_{x \to 5} \frac{x^2 + 25}{x - 5}$

d) $\displaystyle\lim_{x \to 5} \frac{x^2 + 25}{x + 5}$ e) $\displaystyle\lim_{x \to -5} \frac{x - 5}{x^2 - 25}$ f) $\displaystyle\lim_{x \to -5} \frac{x + 5}{x^2 - 25}$

g) $\displaystyle\lim_{x \to -5} \frac{x - 5}{x^2 + 25}$ h) $\displaystyle\lim_{x \to -5} \frac{x + 5}{x^2 + 25}$ i) $\displaystyle\lim_{x \to 5} \frac{x - 5}{x^2 + 25}$

j) $\displaystyle\lim_{x \to 0} \frac{|x| + 1}{x}$ k) $\displaystyle\lim_{x \to 0} \frac{|x|}{x + 1}$ l) $\displaystyle\lim_{x \to 1} \frac{|x|}{x + 1}$

m) $\displaystyle\lim_{x \to 1} \frac{|x|}{x + 1}$ n) $\displaystyle\lim_{x \to 0} \frac{x}{|x| + 1}$

2. Suppose $f : \mathbb{R} \to \mathbb{R}$ is defined as follows:

$$f(x) = \begin{cases} x^2 & \text{if } x \text{ is rational} \\ -x^2 & \text{if } x \text{ is irrational.} \end{cases}$$

Evaluate the following.

a) $\displaystyle\lim_{x \to 0} f(x)$ b) $\displaystyle\lim_{x \to 5} f(x)$ c) $\displaystyle\lim_{x \to \sqrt{2}} f(x)$

3. Suppose $f : \mathbb{R} \to \mathbb{R}$ is defined as follows:

$$f(x) = \begin{cases} x^2 & \text{if } x \text{ is irrational} \\ x & \text{if } x \text{ is rational.} \end{cases}$$

Evaluate the following.

a) $\lim_{x \to 0} f(x)$ b) $\lim_{x \to 5} f(x)$ c) $\lim_{x \to \sqrt{2}} f(x)$

4. Suppose $f : \mathbb{R} \to \mathbb{R}$ is defined as follows:

$$f(x) = \begin{cases} 0 & \text{if } x \text{ is irrational} \\ 1/t & \text{if } x = s/t. \end{cases}$$

where s/t is a fraction in its lowest terms (i.e., $f(24/17) = 1/17$).

Evaluate the following.

a) $\lim_{x \to 0} f(x)$ b) $\lim_{x \to 5} f(x)$ c) $\lim_{x \to \sqrt{2}} f(x)$

5. Give an example of a function that is bounded on $(-2, 3)$ and has no limit at 2.

6. Give an example of a function that is bounded on $(-2, 3)$ and has no limit at either 1 or 2.

7. Give an example of a function that is bounded on $(-2, 3)$ and has no limit at -1, 0, 1, and 2.

8. Show that $\lim_{x \to 0} \sin(1/x)$ does not exist.

9. Prove Theorem 11.2.2.2.

10. Prove Theorem 11.2.2.3.

11. Prove Theorem 11.2.2.4.

12. Prove Proposition 11.2.1.3.

13. Suppose $f, g : D \to \mathbb{R}$ are functions such that $\lim_{x \to a} f(x) = A \neq 0$ and $\lim_{x \to a} g(x) = 0$. Prove that $\lim_{x \to a} f(x)/g(x)$ does not exist.

11.3 CONTINUITY

The function $f : D \to \mathbb{R}$ is said to be *continuous at* $a \in D$ if

$$\lim_{x \to a} f(x) = f(a).$$

It is implicit in this definition that f must be *defined* at a in order for it to be continuous there. The function $f : D \to \mathbb{R}$ is said to be continuous if it is continuous at every number a in its domain D. The continuity of all the functions in the next proposition follows immediately from Proposition 11.2.1.

Proposition 11.3.1 *The constant function $f(x) = r$, the identity function $g(x) = x$, and the absolute value function $h(x) = |x|$ are all continuous.*

The following theorem, which is an analog of Theorems 7.2.3 and 11.2.2 provides us with a host of continuous functions.

Theorem 11.3.2 *Suppose $f, g : D \to \mathbb{R}$ are two functions that are both continuous at $a \in D$. Then*

1. $f(x) + g(x)$ *is continuous at a*

2. $f(x)g(x)$ *is continuous at a*

3. $f(x) - g(x)$ *is continuous at a*

4. $f(x)/g(x)$ *is continuous at a provided $g(a) \neq 0$.*

Proof of 1. By part 1 of Theorem 11.2.2 and the fact that $f(x)$ and $g(x)$ are each continuous, it follows that

$$\lim_{x \to a} (f(x) + g(x)) = \lim_{x \to a} f(x) + \lim_{x \to a} g(x) = f(a) + g(a).$$

Hence, $f(x) + g(x)$ is continuous at x.

Proof of 2. See Exercise 28.

Proof of 3. See Exercise 29.

Proof of 4. See Exercise 30. □

It follows from the several parts of the above theorem that the function

$$f(x) = \frac{x^3 - 3x^2 - x + 2}{x^2 - x}$$

is continuous at each value of x except at 0 and 1, values which are not in its domain.

Let $f : D \to \mathbb{R}$ and $a \in D$. If $\lim_{x \to a} f(x)$ does not exist, or if it exists and does not equal $f(a)$, then the function $f(x)$ is not continuous, or *discontinuous* at a. Thus, since the Dirichlet function (Equation 1) has no limit at any a it is discontinuous at every a. The next proposition says that the composition of continuous functions is also continuous. This proposition is preceded by a lemma whose easy proof is relegated to Exercise 27. Note that the only difference between this lemma and the definition of continuity is that the lemma allows for the sequences in question to contain the limiting value a whereas the definition does not.

Lemma 11.3.3 *Let $f : D \to \mathbb{R}$ be continuous at $a \in D$. If $\{x_n\} \subset D$ and $\{x_n\} \to a$ then $\{f(x_n)\} \to f(a)$.*

Proposition 11.3.4 *Let $f : D \to \mathbb{R}$ and $g : E \to \mathbb{R}$ be two functions such that the composition $f \circ g$ is defined in E. If g is continuous at $a \in E$ and f is continuous at $g(a)$, then $f \circ g$ is also continuous at a.*

Proof. Let $\{x_n\} \subset D - \{a\}$ be such that $\{x_n\} \to a$. It follows from the continuity of g that $\{g(x_n)\} \to g(a)$ and it follows from the continuity of f and Lemma 11.3.3 that $\{f(g(x_n))\} \to f(g(a))$. Thus, $\{(f \circ g)(x_n)\} \to (f \circ g)(a)$ and so $f \circ g$ is indeed continuous at a.   □

We now show that some other common functions enjoy the property of continuity.

Proposition 11.3.5 *The function $f(x) = \sqrt{x}$ is continuous at each $a \geq 0$.*

Proof. This follows immediately from Proposition 7.2.12.   □

In order to prove the continuity of the trigonometric functions it is necessary to establish a strange looking limit that turns out to be extremely useful both here and in the next chapter.

Proposition 11.3.6

1. $|\sin \phi| \leq |\phi|$ *for all ϕ*

2. $\lim\limits_{\phi \to 0} \dfrac{\sin \phi}{\phi} = 1.$

Proof of part 1. Since this is obvious for $\phi = 0$ and $|\sin(-\phi)| = |\sin \phi|$, it follows that we may restrict attention to positive values of ϕ. Moreover, for $\phi \geq \pi/2$ we have

$$\phi \geq \frac{\pi}{2} > 1 \geq \sin \phi,$$

and so we may restrict further to $0 < \phi < \pi/2$. In this range of ϕ, Figure 11.1, wherein $OA = OB = 1$ and $BC \| AD \perp OA$, is applicable and so we conclude that

$$\text{area of } \triangle\, OAB < \text{ area of sector } OAB < \text{ area of } \triangle OAD$$

or

$$\frac{1}{2}(OA \cdot BC) \quad < \quad \pi \cdot 1^2 \cdot \frac{\phi}{2\pi} \quad < \quad \frac{1}{2}(OA \cdot AD)$$

or

$$\frac{\sin \phi}{2} \quad < \quad \frac{\phi}{2} \quad < \quad \frac{\tan \phi}{2} \tag{2}$$

which establishes Inequality 1.

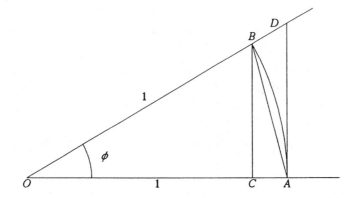

Fig. 11.1   A comparison of an angle to its trigonometric functions.

Proof of part 2. Since

$$\frac{\sin(-\phi)}{-\phi} = \frac{-\sin\phi}{-\phi} = \frac{\sin\phi}{\phi}$$

it may again be assumed that $0 < \phi < \pi/2$. Continuing from Inequality (2),

$$1 < \frac{\phi}{\sin\phi} < \frac{1}{\cos\phi}$$

or

$$1 > \frac{\sin\phi}{\phi} > \cos\phi = \sqrt{1 - \sin^2\phi} \geq \sqrt{1 - \phi^2}.$$

Hence, by Proposition 7.2.5, $\lim_{\phi\to 0} \sin\phi/\phi = 1$.   □

Proposition 11.3.7 *The functions* $\sin x$ *and* $\cos x$ *are continuous at each* x.

Proof. Suppose $\{x_n\} \to A$. It follows from the trigonometric identity

$$\sin\alpha - \sin\beta = 2\sin\frac{\alpha - \beta}{2}\cos\frac{\alpha + \beta}{2}$$

and Proposition 11.3.6.1 that

$$|\sin x_n - \sin A| \leq 2\left|\sin\frac{x_n - A}{2}\right| \leq 2 \cdot \frac{|x_n - A|}{2} = |x_n - A|.$$

Hence, by Proposition 7.2.5, $\{\sin x_n\} \to \sin A$ and so $f(x) = \sin x$ is continuous for each value of x. The proof of the continuity of $\cos x$ is similar and is relegated to Exercise 21.   □

   The continuity of the other trigonometric functions over their domains follows from standard trigonometric identities and Theorem 11.3.2.4.

A function $f : D \to \mathbb{R}$ is said to be an *extension* of the function $g : E \to \mathbb{R}$ if $D \supset E$ and $f(x) = g(x)$ for all $x \in E$. Beginning with a continuous function $f : D \to \mathbb{R}$, it is necessary sometimes to look for a continuous extension $g : E \to \mathbb{R}$. The need for such continuous extensions cannot be motivated within the bounds of this course and we justify their consideration here with the excuse that they generate interesting examples that will hopefully deepen the students' understanding of continuity.

Proposition 11.3.8 *Suppose $a < c < b$ and $f_1 : (a, c) \to \mathbb{R}$ and $f_2 : (c, b) \to \mathbb{R}$ are continuous functions. Then f_1 and f_2 can be extended to a continuous function $f : (a, b) \to \mathbb{R}$ if and only if*

$$\lim_{x \to b} f_1(x) = \lim_{x \to b} f_2(x).$$

Proof. If such an extension f exists, then the equality of the limits follows from the definition of continuity. Conversely, suppose the two limits are equal. Define $f : (a, b) \to \mathbb{R}$ as

$$f(x) = \begin{cases} f_1(x) & \text{for } x \in (a, c) \\ \lim_{x \to b} f_1(x) = \lim_{x \to b} f_2(x) & \text{for } x = c \\ f_2(x) & \text{for } x \in (c, b). \end{cases}$$

The continuity of f for $x \in (a, c) \cup (c, b)$ follows from the continuity of f_1 and f_2. The continuity of f at $x = c$ follows from Proposition 7.2.11. $\square$

Example 11.3.9 The function $f(x) = |x|$ is continuous. The reason for this is that $f(x) = -x$ on $(-\infty, 0)$, $f(x) = x$ on $(0, \infty)$ and

$$\lim_{x \to 0} x = 0 = \lim_{x \to 0} (-x).$$

Example 11.3.10 The function $f(x) = |x|/x$ cannot be continuously extended to all of $\mathbb{R}$. The reason for this is that since

$$\lim_{n \to \infty} \left\{ f\left(\frac{(-1)^n}{n} \right) \right\} = \lim_{n \to \infty} \{(-1)^n\}$$

does not exist, it follows that $\lim_{x \to 0} f(x)$ does not exist either.

Exercises 11.3

1. Explain why the following functions are or are not continuous. Be sure to specify the domain in each case.

a) $f(x) = \dfrac{x + 3}{x^2 - 25}$

b) $f(x) = \dfrac{x+3}{x^2 - 25}$ for $x \neq \pm 5$, $f(\pm 5) = 3$

c) $f(x) = \dfrac{x-3}{x^2 + 25}$

d) $f(x) = \dfrac{x^2 - 25}{x - 5}$

e) $f(x) = \dfrac{x^2 - 25}{x - 5}$ for $x \neq 5$, $f(5) = 1$

f) $f(x) = \dfrac{x^2 - 25}{x - 5}$ for $x \neq 5$, $f(5) = 10$

g) $f(x) = \dfrac{\sin x}{x}$

h) $f(x) = \dfrac{\sin x}{x}$ for $x \neq 0$, $f(0) = 0$

i) $f(x) = \dfrac{\sin x}{x}$ for $x \neq 0$, $f(0) = 1$

j) $f(x) = |x| - 1$

k) $f(x) = \dfrac{|x| - 1}{x}$

l) $f(x) = \dfrac{|x|}{x - 1}$

m) $f(x) = \left| \dfrac{x - 1}{x^2 + 1} \right|$

2. Suppose the function $f : D \to \mathbb{R}$ has the property that for some fixed constant c, $|f(x) - f(y)| < c|x - y|$ for all $x, y \in D$. Prove that f is continuous.

3. Is there a continuous function $f : \mathbb{R} \to \mathbb{R}$ such that $f(x) = \sin x / x$ whenever $x \neq 0$?

4. Is there a continuous function $f : \mathbb{R} \to \mathbb{R}$ such that $f(x) = |x|/x$ whenever $x \neq 0$?

5. Is there a continuous function $f : \mathbb{R} \to \mathbb{R}$ such that $f(x) = \sin(1/x)$ whenever $x \neq 0$?

6. Is there a continuous function $f : \mathbb{R} \to \mathbb{R}$ such that $f(x) = x \sin(1/x)$ whenever $x \neq 0$?

7. Is there a continuous function $f : \mathbb{R} \to \mathbb{R}$ such that $f(x) = x^2 \sin(1/x)$ whenever $x \neq 0$?

8. Is there a continuous function $f : [0, \infty) \to \mathbb{R}$ such that $f(x) = (\sin x)/\sqrt{x}$ whenever $x > 0$?

9. Is there a continuous function $f : [0, \infty) \to \mathbb{R}$ such that $f(x) = \sqrt{x} \sin x$ whenever $x > 0$?

10. Is there a continuous function $f : \mathbb{R} \to \mathbb{R}$ such that $f(x) = x$ for $x > 0$ and $f(x) = x^2$ for $x < 0$?

11. Is there a continuous function $f : \mathbb{R} \to \mathbb{R}$ such that $f(x) = x^2 + 1$ for $x > 0$ and $f(x) = x^2$ for $x < 0$?

12. Is there a continuous function $f : \mathbb{R} \to \mathbb{R}$ such that $f(x) = \cos x$ for $x > 0$ and $f(x) = \sin x$ for $x < 0$?

13. Is there a continuous function $f : \mathbb{R} \to \mathbb{R}$ such that $f(x) = \sin x$ for $x > 0$ and $f(x) = x^2$ for $x < 0$?

Exercises 14 and 15 describe an alternate approach to the definition of continuity. This new definition is essentially the same as that employed by Bolzano.

14. Suppose $f : \mathbb{R} \to \mathbb{R}$ is continuous at a. Prove that given any real number $\varepsilon > 0$, there exists another real number δ (whose value may depend on ε) such that $|f(x) - f(a)| < \varepsilon$ whenever $|x - a| < \delta$.

15. Let $f : \mathbb{R} \to \mathbb{R}$ and let $a \in D$. Suppose that for every real number $\varepsilon > 0$, there exists another real number δ (whose value may depend on ε) such that $|f(x) - f(a)| < \varepsilon$ whenever $|x - a| < \delta$. Prove that f is continuous at a.

16. Suppose $f : [a, b] \to \mathbb{R}$ is continuous. Show that given any real number $\varepsilon > 0$, there exists a real number δ (whose value may depend on ε) such that $|f(x) - f(y)| < \varepsilon$ whenever $|x - y| < \delta$.

17. Suppose $f : \mathbb{R} \to \mathbb{R}$ is defined as follows:

$$f(x) = \begin{cases} 0 & \text{if } x \text{ is irrational} \\ 1/q & \text{if } x = p/q \end{cases}$$

where p/q is a fraction in its lowest terms (i.e., $f(24/17) = 1/17$). Prove that $f(x)$ is continuous for all irrational x and discontinuous for all rational x.

18. Suppose $f : \mathbb{R} \to \mathbb{R}$ is defined as follows:

$$f(x) = \begin{cases} x^2 & \text{if } x \text{ rational} \\ -x^2 & \text{if } x \text{ irrational.} \end{cases}$$

Prove that f is continuous only at 0.

19. Suppose $f : \mathbb{R} \to \mathbb{R}$ is defined as follows:

$$f(x) = \begin{cases} x^2 & \text{if } x \text{ is irrational} \\ x & \text{if } x \text{ is rational.} \end{cases}$$

Where is f continuous and where is it discontinuous?

20. The *greatest integer function* $[x]$ denotes the greatest integer not exceeding x. Thus, $[2.77] = 2$, $[-2.77] = -3$, and $[17] = 17$.

a) Where is the function $f(x) = [x]$ continuous?

b) Where is the function $f(x) = [10x]$ continuous?

c) Where is the function $f(x) = [2x]$ continuous?

d) Where is the function $f(x) = \left[\sqrt{2x}\right]$ continuous?

21. Prove that $f(x) = \cos x$ is a continuous function.

22. Use the previous exercise to decide on the continuity of the following functions.

a) $f(x) = \dfrac{\cos x}{x}$

b) $f(x) = \dfrac{\cos x}{x}$ for $x \neq 0$, $f(0) = 0$

c) $f(x) = \dfrac{\cos x}{x}$ for $x \neq 0$, $f(0) = 1$

23. Prove that $\lim_{x \to 0} \frac{\tan x}{x} = 1$.

24. Prove that $\tan x$ is a continuous function. What is its domain?

25. Use the previous exercises to decide on the continuity of the following functions.

a) $f(x) = \dfrac{\tan x}{x}$ for $x \in (0, \pi/2)$

b) $f(x) = \dfrac{\tan x}{x}$ for $x \in (0, \pi/2)$, $f(0) = 0$

c) $f(x) = \dfrac{\tan x}{x}$ for $x \in (0, \pi/2)$, $f(0) = 1$.

26. Let $f : \mathbb{R} \to \mathbb{R}$ be a continuous function such that $f(x + y) = f(x) + f(y)$ for all x, $y \in \mathbb{R}$. Prove that there exists a number c such that $f(x) = cx$ for all x (Hint: Make use of Exercise 6.5.1).

27. Prove Lemma 11.3.3.

28. Prove Theorem 11.3.2.2.

29. Prove Theorem 11.3.2.3.

30. Prove Theorem 11.3.2.4.

31. Prove that if $f : \mathbb{R} \to \mathbb{R}$ is continuous, then so is the function $|f| : D \in \mathbb{R}$ continuous.

32. Find a discontinuous function $f : D \to \mathbb{R}$ such that $|f| : D \in \mathbb{R}$ is continuous.

33. Suppose $f : \mathbb{R} \to \mathbb{R}$ is continuous and $f(x) = f(x/2)$ for all x. Prove that $f(x)$ is a constant function.

34. Suppose $b \neq 0$, 1 and a are real numbers, and $f : \mathbb{R} \to \mathbb{R}$ is continuous. Prove that if $f(x) = f((x + a)/b)$ for all x, then $f(x)$ is a constant function. Why is this not necesarily true when $b = 1$?

35. Show that if $f : \mathbb{R} \to \mathbb{R}$ is a continuous function such that $f(x) = f(x^2)$ for all x then $f(x)$ is a constant function.

36. Show that if $0 \leq c \leq 1/4$ is a fixed number and $f : \mathbb{R} \to \mathbb{R}$ is a continuous function such that $f(x) = f(x^2 + c)$ for all x, then $f(x)$ is a constant function.

37. Suppose $f : D \to \mathbb{R}$ is continuous and $f(x^*) \neq 0$ for some $x^* \in D$. Prove that there is a neighborhood (a, b) of x^* such that $0 \notin f([a, b])$.

11.4   PROPERTIES OF CONTINUOUS FUNCTIONS

As mentioned in this chapter's introduction, mathematicians applied the word *continuity* to functions long before they agreed on a common formal definition. Regardless of what the elusive definition might turn out to be, they all had some expectations (or should we say preconceptions) regarding the properties such functions must possess. Not surprisingly, a common expectation was that the range of a continuous function on an unbroken domain should have no gaps. The next proposition shows that our definition of continuity fulfills this expectation.

Theorem 11.4.1 (Intermediate Value Theorem) *Let $f : [a, b] \to \mathbb{R}$ be a continuous function and let y^* be a real number such that either $f(a) < y^* < f(b)$ or $f(b) < y^* < f(a)$. Then there exists $x^*, a < x^* < b$, such that $f(x^*) = y^*$.*

Proof. We prove the case $f(a) < y^* < f(b)$ and relegate the other case to Exercise 5. The proof zeros in on x^* by means of a binary search technique. Set $a_0 = a$ and $b_0 = b$. Assuming that $a_n < b_n$ have been defined so that

$$f(a_n) < y^* < f(b_n), \tag{3}$$

define $[a_{n+1}, b_{n+1}]$ to be that half of $[a_n, b_n]$ such that

$$f(a_{n+1}) < y^* < f(b_{n+1}).$$

More formally, set

$$a_{n+1} = a_n, \quad b_{n+1} = \frac{a_n + b_n}{2} \quad \text{if } f\left(\frac{a_n + b_n}{2}\right) > y^*$$

$$a_{n+1} = \frac{a_n + b_n}{2}, \quad b_{n+1} = b_n \quad \text{if } f\left(\frac{a_n + b_n}{2}\right) < y^*$$

$$x^* = \frac{a_n + b_n}{2} \quad \text{if } f\left(\frac{a_n + b_n}{2}\right) = y^* \tag{4}$$

It is clear that if Equation (4) ever occurs then we are done, and so we assume that it does not happen, in which case this process defines two bounded monotone sequences $\{a_n\}$, $\{b_n\}$ such that for each n

$$b_n - a_n = \frac{b_{n-1} - a_{n-1}}{2} = \frac{b_{n-2} - a_{n-2}}{2^2} = \ldots = \frac{b_0 - a_0}{2^n}.$$

It follows from Proposition 8.1.4 that both sequences $\{a_n\}$ and $\{b_n\}$ converge to the same limit, say x^*. The continuity of f guarantees that the sequences $\{f(a_n)\}$ and $\{f(b_n)\}$ both converge to $f(x^*)$. It follows from (3) and Proposition 7.2.10 that

$$f(x^*) = \lim_{n \to \infty} \{f(a_n)\} \le y^* \le \lim_{n \to \infty} \{f(b_n)\} = f(x^*).$$

Hence, $f(x^*) = y^*$. $\qquad\qquad\qquad\qquad\qquad\qquad\qquad\qquad\qquad\qquad$ □

Example 11.4.2 Consider the function $f(x) = x^3 - 2x - 5$. Since

$$f(2) = -1 < 0 < 16 = f(3)$$

it follows from the Intermediate Value Theorem that there is a number r between 2 and 3 such that

$$f(r) = r^3 - 2r - 5 = 0.$$

Since $f(2.5) = 5.625 > 0$, it follows that $2 < r < 2.5$. Since $f(2.25) = 1.890\ldots$, it follows that $2 < r < 2.25$. Iterations of this procedure will pinpoint r to any desired accuracy. This method can be clearly used to solve any equation that has solutions. It is, however, highly inefficient, and other methods are used in practice.

The first rigorous proof of the Intermediate Value Theorem was given by Bolzano in 1817. It was in this connection that he first stated, as a theorem whose proof was based on our Theorem 9.3.1 (which both he and Cauchy took for granted), the Axiom of Completeness. His statement was

If a property M does not belong to *all* values of a variable x, but does belong to *all* values which are *less* than a certain u, then there is always a quantity U which is the greatest of those of which it can be asserted that all smaller x have property M.

To see the relationship of Bolzano's statement to our Completeness Axiom, let S denote the set of all those numbers u such that all x less than u have the property M. Then the least upper bound of S is Bolzano's U (see Exercise 6.3.7).

As yet another illustration of the power of the Intermediate Value Theorem we offer the following considerably easier alternate proof of the existence of the root $\sqrt[n]{r}$ (see Proposition 6.3.4).

Corollary 11.4.3 *For every positive integer n and positive real number r there exists a real number x such that $x^n = r$.*

Proof. Set $f(x) = x^n$, $y^* = r$, $a = 0$ and $b = 1 + r$. Then

$$f(a) = 0 < y^* = r < 1 + r < (1 + r)^n = f(b).$$

It follows from the Intermediate Value Theorem that there exists x^* such that

$$x^{*^n} = f(x^*) = y^* = r. \qquad \square$$

It follows from Proposition 6.2.2.10 that the number x^* whose existence is guaranteed by the above proposition is in fact unique. Thus the function $x^{1/n} = \sqrt[n]{x}$ is well defined.

Exercise 10 makes it clear that the intermediate value property is not synonymous with continuity. In other words, there are discontinuous functions that also possess the intermediate value property.

A major portion of first year calculus is devoted to the computation of maxima and minima, and certainly its success in locating these critical points is impressive. The following shows that the existence of maxima and minima is dependent on continuity alone, and has little to do with differentiation.

The function $f : D \to \mathbb{R}$ is said to assume a maximum (or minimum) in D if there exists a number x_M (or x_m) in $[a, b]$ such that

$$f(x_M) \geq f(x) \quad (\text{or } f(x_m) \leq f(x)) \quad \text{for all } x \in D.$$

Thus the function $f(x) = -x^2$ assumes a maximum in $\mathbb{R}$ (at $x_M = 0$) but does not assume a minimum. The function $f(x) = x$ assumes neither a maximum nor a minimum in $(0, 1)$ but does assume both in $[0, 1]$ (at $x_M = 1$ and $x_m = 0$ respectively). The function

$$f(x) = \begin{cases} x & \text{for } 0 \leq x < 1 \\ -x & \text{for } 1 \leq x \leq 2 \end{cases}$$

assumes a minimum in $[0, 2]$ (at $x = 2$), but does not assume a maximum in that interval. The following fundamental result was first proven by Karl Weierstrass (1815–1897) in 1861.

Theorem 11.4.4 (The Maximum Principle) *Suppose $f : [a, b] \to \mathbb{R}$ is a continuous function. Then f assumes both a maximum and a minimum in $[a, b]$.*

Proof. Only the assumption of a maximum is proved, leaving the proof for the minimum to Exercise 11. First it is shown that f is bounded on $[a, b]$. Suppose not: then for each positive integer n there exists a number $x_n \in [a, b]$ such that $|f(x_n)| > n$. Since the sequence $\{x_n\}$ is itself bounded between a and b, it follows from the Bolzano–Weierstrass Theorem that it has a convergent subsequence $\{x_{n_k}\}$. The continuity of f implies that $\{f(x_{n_k})\}$ is also convergent, thus contradicting the fact that since $|f(x_{n_k})| > n_k$ this sequence is unbounded.

Now that we know that f is bounded on $[a, b]$, it follows from the Completeness Axiom that $f([a, b])$ has a least upper bound, say L. For any positive integer n, $L - 1/n$ is not an upper bound of $f([a, b])$ and so there exists a number $\bar{x}_n \in [a, b]$ such that

$$L - \frac{1}{n} < f(\bar{x}_n) \le L.$$

It follows that $\{f(\bar{x}_n)\} \to L$. By the Bolzano–Weierstrass Theorem there exists a convergent subsequence $\{\bar{x}_{n_k}\}$ of $\{\bar{x}_n\}$, say $\lim_{k\to\infty}\{\bar{x}_{n_k}\} = \bar{x} \in [a, b]$. Since $\{f(\bar{x}_{n_k})\}$ is a subsequence of $\{f(\bar{x}_n)\} \to L$ we have

$$\lim_{k \to \infty} f(\bar{x}_{n_k}) = L.$$

On the other hand, the continuity of f implies that

$$\lim_{k \to \infty} f(\bar{x}_{n_k}) = f(\bar{x}).$$

Thus, $f(\bar{x}) = L$ and so f assumes its maximum at $\bar{x}$. $\square$

Example 11.4.5 The function

$$f(x) = \frac{\sin x + \cos^2 x}{1 + |x| + |x - 2| + x^{16}}$$

is clearly continuous everywhere. Consequently, it has a maximum in any interval whatsoever. Locating this maximum, however, is another story, one which does not belong in this text.

Exercises 11.4

1. Prove that if $f : [0, 1] \to [0, 1]$ is a continuous function, then there exists a number $x^* \in [0, 1]$ such that $f(x^*) = x^*$.

2. Prove that if $f : [a, b] \to [a, b]$ is a continuous function, then there exists a number $x^* \in [a, b]$ such that $f(x^*) = x^*$.

3. Let $f : [0, 2] \to \mathbb{R}$ be a continuous function such that $f(0) = f(2)$. Prove that there exists a number $x^* \in [0, 1]$ such that $f(x^*) = f(x^* + 1)$. (Hint: Consider the function $g(x) = f(x) - f(x + 1)$.)

4. Prove that a continuous function that assumes only irrational values is necessarily constant.

5. Complete the proof of Theorem 11.4.1 for the case $f(a) > f(b)$.

6. Prove that if $f(x)$ is a polynomial of odd degree, then there exists a real number x^* such that $f(x^*) = 0$.

Problems 7–9 are to be done without a calculator.

7. Prove that the equation $x^4 - 5x^3 + x - 1 = 0$ has a solution.

8. Prove that the equation $\sin x = x - 1$ has a solution.

9. Prove that the equation $2 \sin x = \cos x + 1 - x^2$ has a solution.

10. Let $f(x) = \sin(1/x)$ for $x \neq 0$ and $f(0) = 0$.

   a) Show that this function is not continuous at 0;
   b) Show that this function does possess the intermediate value property. In other words, show that if $a < b$ and y^* is any number between $f(a)$ and $f(b)$, then there exists a number x^*, $a < x^* < b$ such that $f(x^*) = y^*$.

11. Complete the proof of Theorem 11.4.4 by showing that $f(x)$ also assumes its minimum.

12. Prove that the Intermediate Value Theorem remains valid when the domain is changed from $[a, b]$ to (a, b).

13. Prove that if $f : [a, b] \to \mathbb{R}$ is a continuous function then $f([a, b]) = [c, d]$ for some $c, d \in \mathbb{R}$. (Hint: Set $c = \min f([a, b])$ and $d = \max f([a, b])$.)

14. Prove that if $f : [a, b] \to \mathbb{R}$ is continuous and one to one then it is monotone.

15. Prove that there exists no continuous function $f : \mathbb{R} \to \mathbb{R}$ that assumes each of its values exactly twice.

16. Describe the main mathematical achievements of
   a) Bernard Bolzano
   b) Karl Weierstrass

Chapter Summary

In this chapter, a discussion of the evolution of the notion of continuity of functions is followed by its rigorous definition in terms of limits of functions which, in turn, are defined by means of limits of sequences. A variety of propositions are proved that imply the continuity of the standard rational and trigonometric functions. Two important implications of the continuity of a function are also proved, these are the Intermediate Value Theorem and the Maximum Principle.

12

Differentiability

This chapter covers differentiability, its consequences, and the indefinite integral.

12.1 AN INFORMAL INTRODUCTION TO DIFFERENTIATION

Recall that Fermat's max/min method of Chapter 2 made use of a procedure that greatly resembled modern day differentiation. A similar method was used by Fermat by 1637 for the purpose of constructing tangents to given curves. Thus, it would not be improper to state that Fermat could differentiate some functions. Fermat's method employed a quantity e that was mysteriously both different from and equal to 0. Descartes developed similar techniques at about the same time.

The differentiation process that Newton described in *De Analysi* made use of *moments* which later became known as *infinitesimals*. Like Fermat's e, the values of these quantities seemed to be alternating between zero and non-zero, depending mostly on convenience. This approach was later abandoned by Newton in favor of *fluxions*. Here the value of each variable was stipulated to be actually changing, that is in a state of flux, and its fluxion was that variable's rate of change. The variable x's fluxion was denoted by $\dot{x}$ and equaled, in modern notation, dx/dt. If the variables x and y were related, then the derivative of y with respect to x was represented by Newton as the ratio $\dot{y}/\dot{x}$ which, of course equals

$$\frac{dy/dt}{dx/dt} = \frac{dy}{dx}$$

When Leibniz developed his version of calculus in the 1680's, he created the now familiar *differentials* $dx, dy, \ldots$, although he did refer to them as *differences*. It is not clear what meaning he attached to these quantities. His diagrams indicate that, at least sometimes, he thought of these as fixed quantities. On the other hand, *Leibniz's Power Rule*

$$dx^a = ax^{a-1}dx$$

makes it clear that at other times dx was a quantity that approached zero, and so, of necessity, a variable quantity.

In 1734 Bishop Berkeley, fed up with the disdain with which some scientists regarded religion, published a tract *The Analyst* in which he counterattacked them by criticizing the supposedly irreproachable logic of mathematics. Being a competent mathematician himself, he knew where to find the weak spots, namely in the foundations of calculus. "He who can digest a second or third fluxion. . .," wrote Berkeley, "need not, methinks, be squeamish about any point in theology." His apt charaterization of infinitesimals as "ghosts of departed quantities" must also have stung.

Unsuccessful efforts to rigorize calculus continued into the nineteenth century. Curiously, Euler's book *Introduction to Analysis of the Infinite* makes no mention of the differential calculus. Instead, it is devoted almost exclusively to infinite series and their applications to algebra and geometry. Elsewhere, though, Euler (1755) made it clear that his view of differentials agreed with that of Leibniz. Lagrange, in his 1797 book, in order to dispense with the troublesome infinitesimals, used infinite series to <u>define</u> derivatives. Specifically, he assumed (wrongly, as it turns out) that given a function $f(x)$ and a quantity h, there exist functions $c_1(x), c_2(x), c_3(x), \ldots$ such that

$$f(x+h) = f(x) + c_1(x)h + c_2(x)h^2 + c_3(x)h^3 + \ldots$$

(Readers may recognize this as the Taylor series of $f(x)$.) Lagrange then defined $c_1(x)$ to be the derivative of $f(x)$. He actually referred to the functions $c_1(x), c_2(x), c_3(x), \ldots$ as the derived functions of the primitive function $f(x)$, from which the term derivative is descended.

It is from Cauchy's *Cours d'Analyse* of 1821 that comes the modern definition of the derivative as the limit of a ratio of the form

$$\frac{f(x+\alpha) - f(x)}{\alpha}$$

in which the denominator α converges to 0. While Cauchy still made occasional use of the term *infinitesimal* in describing the behavior of functions, his understanding of this term was different from those of Fermat and Leibniz. To Cauchy an infinitesimal was a variable that converged (through <u>successive</u> numerical values) to 0, as opposed to a quantity that is sometimes zero and sometimes not. His employment of the word <u>successive</u> in describing the way an infinitesimal varies would seem to imply that Cauchy visualized this infinitesimal as a <u>sequence</u> that converged to 0.

12.2   THE DERIVATIVE

Given a function $f : D \to \mathbb{R}$ and a number $a \in D$, we define the *derivative* of f at a to be the value of

$$\lim_{h \to 0} \frac{f(a+h) - f(a)}{h} = \lim_{x \to a} \frac{f(x) - f(a)}{x - a}$$

whenever this limit exists. Should this limit exist for a specific a, the function f is then said to be *differentiable* at a and the value of the derivative is denoted by $f'(a)$. If E denotes the set of a's at which the function f is differentiable, then $f' : E \to \mathbb{R}$ is the function that assigns to each $a \in E$ the value $f'(a)$. The function f' is also called the *derivative* of f. If $E = \mathbb{R}$ then f is said to be *everywhere differentiable*.

Proposition 12.2.1 *The constant function $f(x) = r$ and the identity function $g(x) = x$ are both everywhere differentiable. In fact, $(r)' = 0$ and $(x)' = 1$.*

Proof. If f is the constant function $f(x) = r$, then, for every number a

$$\lim_{h \to 0} \frac{f(a+h) - f(a)}{h} = \lim_{h \to 0} \frac{r - r}{h}$$
$$= \lim_{h \to 0} \frac{0}{h} = \lim_{h \to 0} 0 = 0.$$

Thus, the constant function $f(x) = r$ is everywhere differentiable and $f'(x) = 0$ for all x.

If g is the identity function $g(x) = x$, then

$$\lim_{h \to 0} \frac{g(a+h) - g(a)}{h} = \lim_{h \to 0} \frac{(a+h) - a}{h}$$
$$= \lim_{h \to 0} \frac{h}{h} = \lim_{h \to 0} 1 = 1.$$

Thus, the identity function $g(x) = x$ is everywhere differentiable and $g'(x) = 1$ for all x. $\square$

Theorem 12.2.5 provides many differentiable functions. First, however, it is necessary to clarify the relationship between differentiability and continuity. The next example demonstrates that the two concepts are not synonymous.

Example 12.2.2 The absolute value function $g(x) = |x|$ is known to be continuous. Nevertheless, it is not differentiable at $x = 0$ since

$$\lim_{n \to \infty} \frac{g\left(\frac{1}{n}\right) - 0}{\frac{1}{n} - 0} = \lim_{n \to \infty} \frac{\frac{1}{n} - 0}{\frac{1}{n} - 0} = 1$$

whereas

$$\lim_{n \to \infty} \frac{g\left(-\frac{1}{n}\right) - 0}{-\frac{1}{n} - 0} = \lim_{n \to \infty} \frac{\frac{1}{n} - 0}{-\frac{1}{n} - 0} = -1 \neq 1.$$

First, it must be proven that differentiability implies continuity. The following lemma, stated explicitly by Lagrange in 1797, is very helpful for this as well as other purposes. His proof, however, was based on the invalid assumption that all functions have power series expansions.

Lemma 12.2.3 *If the function* $f : D \to \mathbb{R}$ *is differentiable at* $a \in D$ *then there exists a function* $e : D \to \mathbb{R}$ *such that*

1. $e(x)$ *is continuous at* a *and* $e(a) = f'(a)$

2. $f(x) = f(a) + (x - a) \cdot e(x)$

Proof. Set

$$e(x) = \frac{f(x) - f(a)}{x - a} \qquad \text{for } x \in D - \{a\} \text{ and } e(a) = f'(a)$$

Equation 2 is clearly satisfied. Moreover, setting $h = x - a$ we see that

$$\lim_{x \to a} e(x) = \lim_{h \to 0} \frac{f(a + h) - f(a)}{h} = f'(a) = e(a),$$

so that Equation 1 is satisfied as well.   □

Exercise 27 contains a converse of this lemma.

Proposition 12.2.4 *If the function* $f : D \to \mathbb{R}$ *is differentiable at* a *then it is also continuous at* a.

Proof. Suppose $f : D \to \mathbb{R}$ is differentiable at a, and let $e : D \to \mathbb{R}$ be the function whose existence is guaranteed by Lemma 12.2.3. It follows from the fact that

$$f(x) = f(a) + (x - a) \cdot e(x)$$

and the continuity of $e(x)$ at a that $\lim_{x \to a} f(x) = f(a)$.   □

Example 12.2.2 of course illustrates the fact that the converse relation does not hold. Continuity does not entail differentiability. The next theorem proves the differentiability of many functions and its proof makes repeated use of Proposition 12.2.4.

Theorem 12.2.5 *Let f, $g : D \to \mathbb{R}$ be differentiable at $a \in D$. Then*

1. $(f + g)'(a) = f'(a) + g'(a)$

2. $(f \cdot g)'(a) = f'(a)g(a) + f(a)g'(a)$

3. $(f - g)'(a) = f'(a) - g'(a)$

4. $\left(\dfrac{f}{g}\right)'(a) = \dfrac{f'(a)g(a) - f(a)g'(a)}{g^2(a)}$ *provided $g(a) \neq 0$.*

Proof of 1.

$$
\begin{aligned}
(f + g)'(a) &= \lim_{h \to 0} \frac{(f + g)(a + h) - (f + g)(a)}{h} \\
&= \lim_{h \to 0} \frac{f(a + h) + g(a + h) - f(a) - g(a)}{h} \\
&\quad \lim_{h \to 0} \frac{f(a + h) - f(a)}{h} + \lim_{h \to 0} \frac{g(a + h) - g(a)}{h} \\
&= f'(a) + g'(a).
\end{aligned}
$$

Proof of 2.

$$
\begin{aligned}
(f \cdot g)'(a) &= \lim_{h \to 0} \frac{(f \cdot g)(a + h) - (f \cdot g)(a)}{h} \\
&= \lim_{h \to 0} \frac{(f(a + h) - f(a)) \cdot g(a + h) + f(a) \cdot (g(a + h) - g(a))}{h} \\
&= \lim_{h \to 0} \frac{f(a + h) - f(a)}{h} \cdot \lim_{h \to 0} g(a + h) \\
&\quad + f(a) \cdot \lim_{h \to 0} \frac{g(a + h) - g(a)}{h} \\
&= f'(a)g(a) + f(a)g'(a).
\end{aligned}
$$

Proof of 3. Since $f - g = f + (-1)g$, this part follows from the previous ones.

Proof of 4. Assume first that f is the constant function $f(x) = 1$. Then,

$$
\begin{aligned}
\left(\frac{1}{g}\right)'(a) &= \lim_{h \to 0} \frac{\dfrac{1}{g(a + h)} - \dfrac{1}{g(a)}}{h} \\
&= \lim_{h \to 0} \frac{g(a) - g(a + h)}{g(a + h) \cdot g(a) \cdot h}
\end{aligned}
$$

$$= \lim_{h \to 0} -\frac{g(a+h) - g(a)}{h} \cdot \lim_{h \to 0} \frac{1}{g(a+h) \cdot g(a)} = \frac{-g'(a)}{g^2(a)}.$$

Applying 2 to arbitrary f/g we have

$$\left(\frac{f}{g}\right)'(a) = \left(f \cdot \frac{1}{g}\right)'(a) = f'(a) \cdot \frac{1}{g(a)} + f(a) \cdot \left(\frac{1}{g}\right)'(a)$$

$$= \frac{f'(a)}{g(a)} - \frac{f(a)g'(a)}{g^2(a)} = \frac{f'(a)g(a) - f(a)g'(a)}{g^2(a)}. \qquad \square$$

It follows from a simple induction argument that $(x^n)' = nx^{n-1}$ whenever n is a positive integer (Exercise 1). Equation 4 of the above theorem then makes it possible to extend this well known formula to negative integers as well. Fractional exponents will be dealt with below, following Proposition 12.2.6. These observations provide us with a host of differentiable functions. The next proposition is known as the *Chain Rule*.

Proposition 12.2.6 *Let $f : D \to \mathbb{R}$ and $g : E \to \mathbb{R}$ be functions such that $f \circ g$ is defined on E. If g is differentiable at b and f is differentiable at $a = g(b)$, then $f \circ g$ is differentiable at b and*

$$(f \circ g)'(b) = f'(a) \cdot g'(b).$$

Proof. It follows from Lemma 12.2.3 that there exists a function $e : D \to \mathbb{R}$, continuous at a, such that

$$f(x) = f(a) + (x - a) \cdot e(x) \quad \text{and} \quad e(a) = f'(a).$$

Consequently,

$$\lim_{h \to 0} \frac{(f \circ g)(b+h) - (f \circ g)(b)}{h}$$

$$= \lim_{h \to 0} \frac{f(g(b+h)) - f(g(b))}{h}$$

$$= \lim_{h \to 0} \frac{f(a) + (g(b+h) - a) \cdot e(g(b+h)) - f(a)}{h}$$

$$= \lim_{h \to 0} \frac{g(b+h) - g(b)}{h} \cdot \lim_{h \to 0} e(g(b+h)) = g'(b) \cdot e(g(b))$$

$$= g'(b) \cdot f'(a)$$

where the penultimate equality is justified by Proposition 11.3.4. $\qquad \square$

Newton's Rule 1 (see Section 3.2) can now be proved.

Proposition 12.2.7 *If $f(x) = x^r$ where r is any rational number, then $f'(x) = rx^{r-1}$.*

Proof. We prove this only for the case $r = 1/n$ where n is a positive integer. The other values of r are relegated to Exercises 1–3.

It follows from Propositions 1.1.2 and 7.2.12 that

$$f'(a) = \lim_{x \to a} \frac{x^{1/n} - a^{1/n}}{x - a}$$

$$= \lim_{x \to a} \frac{1}{x^{(n-1)/n} + x^{(n-2)/n}a^{1/n} + \ldots + x^{1/n}a^{(n-2)/n} + a^{(n-1)/n}}$$

$$= \frac{1}{a^{(n-1)/n} + a^{(n-2)/n}a^{1/n} + \ldots + a^{1/n}a^{(n-2)/n} + a^{(n-1)/n}}$$

$$= \frac{1}{na^{(n-1)/n}} = \frac{1}{n} \cdot a^{(1/n)-1}. \qquad \square$$

The trigonometric functions are also differentiable. This is proved here for the sin x function. The other functions are the subjects of Exercises 4, 5.

Proposition 12.2.8 $(\sin x)' = \cos x.$

Proof. Making use of the trigonometric formula

$$\sin \alpha - \sin \beta = 2 \cos \frac{\alpha + \beta}{2} \sin \frac{\alpha - \beta}{2}$$

we get

$$\lim_{h \to 0} \frac{\sin(a + h) - \sin a}{h} = \lim_{h \to 0} \frac{2 \cos \left(a + \dfrac{h}{2} \right) \sin \dfrac{h}{2}}{h}$$

$$= \lim_{h \to 0} \cos \left(a + \frac{h}{2} \right) \cdot \lim_{h \to 0} \frac{\sin \dfrac{h}{2}}{\dfrac{h}{2}} = (\cos a) \cdot (1) = a,$$

where the last two limits were evaluated by the continuity of the cosine function and Proposition 11.3.6. $\qquad \square$

Exercises 12.2

1. Prove that if $f(x) = x^n$ where n is a positive integer, then $f'(x) = nx^{n-1}$. (Hint: Use mathematical induction and the Product Rule).

2. Prove that if $f(x) = x^{-n}$ where n is a positive integer, then $f'(x) = -nx^{n-1}$.

3. Complete the proof of Proposition 12.2.7.

4. Prove that $(\cos x)' = -\sin x.$

5. Prove that $(\tan x)' = \sec^2 x.$

6. Suppose $f : [a, b] \to \mathbb{R}$ and $g : [b, c] \to \mathbb{R}$ are both differentiable functions. Suppose further that $h(x) = f(x)$ for $a \le x < b$ and $h(x) = g(x)$ for $b \le x \le c$. Prove that $h : [a, c] \to \mathbb{R}$ is differentiable if and only if $f'(a) = g'(a)$.

7. Prove that $f(x) = |x^3|$ is differentiable at 0.

8. Suppose $f(x) = 0$ if x is irrational and $f(x) = 1$ if x is rational. Is f differentiable at 0?

9. Suppose $f(x) = 0$ if x is irrational and $f(x) = x$ if x is rational. Is f differentiable at 0?

10. Suppose $f(x) = 0$ if x is irrational and $f(x) = x^2$ if x is rational. Is f differentiable at 0?

11. Suppose $f(x) = x$ if x is irrational and $f(x) = x^2$ if x is rational. Is f differentiable at 0?

12. Suppose $f(x) = x^2$ if x is irrational and $f(x) = x^3$ if x is rational. Is f differentiable at 0?

13. Suppose $f(x) = \sin(1/x)$ for $x \ne 0$ and $f(0) = 0$. Is f differentiable at 0?

14. Suppose $f(x) = x \sin(1/x)$ for $x \ne 0$ and $f(0) = 0$. Is f differentiable at 0?

15. Suppose $f(x) = x^2 \sin(1/x)$ for $x \ne 0$ and $f(0) = 0$. Is f differentiable at 0?

16. Give an example of a function $f : [a, b] \to \mathbb{R}$ that is differentiable on (a, b) but not continuous on $[a, b]$.

17. Prove that if $f : (a, b) \to \mathbb{R}$ is increasing then $f'(x) \ge 0$ for all $x \in (a, b)$.

18. Prove that if $f : (a, b) \to \mathbb{R}$ is decreasing then $f'(x) \le 0$ for all $x \in (a, b)$.

19. For $f(x) = x^3$ and $a = 3$, find a function $e(x)$ that satisfies the conditions of Lemma 12.2.3.

20. For $f(x) = \sin x^2$ and $a = 0$, find a function $e(x)$ that satisfies the conditions of Lemma 12.2.3.

21. Prove that if $f : \mathbb{R} \to \mathbb{R}$ is differentiable at 0, and if a, b, c are any three real numbers such that $c \ne 0$, then

$$\lim_{x \to 0} \frac{f(ax) - f(bx)}{cx} = \frac{a - b}{c} \cdot f'(0).$$

22. Suppose $f : \mathbb{R} \to \mathbb{R}$ is differentiable at 0. Compute

$$\lim_{x \to 0} \frac{f(x^2) - f(0)}{x}.$$

23. Suppose $f : \mathbb{R} \to \mathbb{R}$ is differentiable at x.

Prove that

$$\lim_{h \to 0} \frac{f(a+h) - f(a-h)}{2h} = f'(a).$$

Give an example of a function $f : \mathbb{R} \to \mathbb{R}$, not differentiable at some a, for which this limit nevertheless exists.

24. Prove that the function $f(x) = x^2 \sin(1/x)$ for $x \neq 0$ and $f(0) = 0$ is differentiable everywhere. Is $f'(x)$ everywhere continuous?

25. Suppose that for some real number c, the function $f : (a, b) \to \mathbb{R}$ satisfies the condition $|f(x) - f(y)| \leq c(x-y)^2$ for all $x, y \in (a, b)$. Prove that $f'(x) = 0$ for each x in (a, b).

26. Show that the function $f(x) = \sqrt[3]{x}$ is not differentiable at $x = 0$, but is continuous there.

27. Let $f : D \to \mathbb{R}$ and $a \in D$. Suppose there exists a real number A and a function $e : D \to \mathbb{R}$ such that $e(x)$ is continuous at a and $f(x) = A + (x - a) \cdot e(x)$. Prove that $f'(a) = e(a)$.

28. Prove that the function $f(x) = |x|/x$ for $x \neq 0$, $f(0) = 1$ for $x = 0$, is not differentiable at 0.

29. Prove that if $f : D \to \mathbb{R}$ is continuous at $a \in D$, and $F(x) = (x - a)f(x)$, then $F'(a) = f(a)$.

30. Show by means of an example that the conclusion of Exercise 29 need not hold if f is not continuous at a.

31. Prove that if $f(x) = x|x|$ then $f'(x) = |x|$.

12.3   THE CONSEQUENCES OF DIFFERENTIABILITY

The tools for proving many of the useful properties of the derivative are now covered. The next theorem should be viewed as a lemma leading to the Mean Value Theorem below.

Theorem 12.3.1 (Rolle's Theorem) *Suppose the function $f : [a, b] \to \mathbb{R}$ is continuous and $f(a) = f(b) = 0$. Suppose further that f is differentiable at each $x \in (a, b)$. Then there exists a real number c, $a < c < b$ such that $f'(c) = 0$.*

Proof. It follows from the Maximum Principle that there exist numbers x_m, $x_M \in [a, b]$ such that

$$f(x_m) \leq f(x) \leq f(x_M) \qquad \text{for all } x \in [a, b]. \tag{1}$$

If $f(x_m) = f(x_M) = 0$, then $f(x)$ has the constant value 0 on $[a, b]$, and so, by Proposition 12.2.1, $f'(c) = 0$ for every c in (a, b). If $f(x_m) \neq 0$ then

$x_m \neq a, b$ and so $f'(x_m)$ exists. In other words,

$$\lim_{h \to 0} \frac{f(x_m + h) - f(x_m)}{h} \tag{2}$$

exists. However, it follows from (12.1) that $f(x_m + h) - f(x_m) \geq 0$ for all h and hence

$$\begin{cases} \dfrac{f(x_m + h) - f(x_m)}{h} \leq 0 & \text{whenever } h < 0 \\ \dfrac{f(x_m + h) - f(x_m)}{h} \geq 0 & \text{whenever } h > 0. \end{cases}$$

Hence, by Proposition 7.2.10, the limit in (2) is both non-positive and non-negative. In other words, $0 \geq f'(x_m) \geq 0$ and so $f'(x_m) = 0$.

The proof of the remaining case where $f(x_M) \neq 0$ is relegated to Exercise 1. $\qquad\square$

The hypotheses of Rolle's Theorem, continuity on $[a, b]$ and differentiability on (a, b), may seem awkward. Nevertheless, there are reasonable functions for which they are necessary. Such, for example, is the function $f(x) = \sqrt{1 - x^2}$ which is continuous on $[-1, 1]$, differentiable on $(-1, 1)$ but not differentiable at $x = \pm 1$ (see Exercise 19).

The reader may recall that many of the applications of the derivative are based on the dubious equations

$$\frac{\Delta y}{\Delta x} = f'(x) \quad \text{or} \quad \Delta y = f'(x)\Delta x.$$

The following theorem is an attempt to rigorize the intuition behind these inaccurate statements. It is due to Cauchy but has its origins in Lagrange's attempt to rigorize calculus. It has a great variety of uses some of which will be demonstrated in several subsequent examples.

Theorem 12.3.2 (The Mean Value Theorem) *Suppose the function $f :$ $[a, b] \to \mathbb{R}$ is continuous and is also differentiable on (a, b). Then there is a number c in (a, b) such that*

$$f'(c) = \frac{f(b) - f(a)}{b - a}.$$

Proof. Define a new function $g : [a, b] \to \mathbb{R}$ via

$$g(x) = f(x) - f(a) - \frac{f(b) - f(a)}{b - a}(x - a).$$

Then g is continuous on $[a, b]$, differentiable in (a, b), and $g(a) = g(b) = 0$. It follows from Rolle's Theorem that there is a number $c \in (a, b)$ such that

$$0 = g'(c) = f'(c) - 0 - \frac{f(b) - f(a)}{b - a} \cdot 1$$

or

$$f'(c) = \frac{f(b) - f(a)}{b - a}. \qquad \square$$

Informally speaking, the above theorem states that if AB is the chord of some smooth and continuous arc, then there is a point on the arc at which the tangent to the arc is parallel to the chord AB (see Fig. 12.1). As recognized by

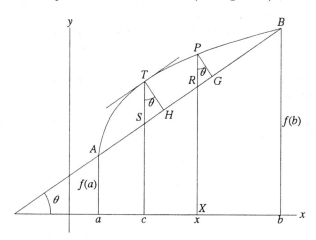

Fig. 12.1   The Mean Value Theorem.

Archimedes (Exercise 1.2.3), this point T is characterized by the fact that of all the points on the arc, it has the greatest distance TH from the chord. Since for arbitrary points P this distance PG is proportional to the vertical segment PR (in fact, $PG = PR \cos \theta$, where θ is the inclination of the chord AB to the positive x–axis), it follows that T also maximizes the vertical segment PR. This line segment PR, whose value depends on x can be expressed in terms of x as follows. The line segment RX is the ordinate of the general point on the straight line AB. It follows from the slope-point form of the equation of a straight line that

$$RX = f(a) + \frac{f(b) - f(a)}{b - a}(x - a).$$

Since PX is the ordinate of the general point on the graph of $y = f(x)$, it follows that $PX = f(x)$ and so

$$PR = PX - RX = f(x) - f(a) - \frac{f(b) - f(a)}{b - a}(x - a)$$

which will be recognized as the function $g(x)$ of the above proof.

The Mean Value Theorem offers us a way of formalizing an approximation that has many uses in calculus, namely, the equation

$$\frac{\Delta y}{\Delta x} \approx f'(x). \qquad (3)$$

If the quantities $b - a$ and $f(b) - f(a)$ of the statement of the Mean Value Theorem are replaced by Δx and Δy respectively, then its conclusion becomes

$$\frac{\Delta y}{\Delta x} = f'(c) \qquad a < c < b. \tag{4}$$

Thus, the Mean Value Theorem turns the vague and approximate equation of (3) into the exact equation of (4).

The next corollary demonstrates the power of Mean Value Theorem in rigorously proving facts assumed in first year calculus.

Corollary 12.3.3 *If the function* $f : [a, b] \to \mathbb{R}$ *has derivative* 0 *at each* x *then it is constant on* $[a, b]$.

Proof. Suppose $x_1, x_2 \in [a, b]$ with $x_1 < x_2$. Then, by the Mean Value Theorem, there exists a number c such that

$$\frac{f(x_2) - f(x_1)}{x_2 - x_1} = f'(c) = 0.$$

It follows that $f(x_1) = f(x_2)$ and so f is indeed constant on $[a, b]$. $\qquad \square$

Corollary 12.3.4 *If* $f : [a, b] \to \mathbb{R}$ *is differentiable and* $f'(x) > 0$ *for all* $x \in (a, b)$, *then* f *is strictly increasing.*

Proof. If $x_1, x_2 \in [a, b]$ such that $x_1 < x_2$, then, by the Mean Value Theorem, there exists a number c such that

$$\frac{f(x_2) - f(x_1)}{x_2 - x_1} = f'(c) > 0.$$

It follows that $f(x_2) > f(x_1)$. $\qquad \square$

The Mean Value Theorem can also be used to prove interesting inequalities.

Example 12.3.5 Prove that if $x > 0$ then $\sqrt{1 + x} < 1 + x/2$.

The function $f(x) = \sqrt{1 + x}$ is continuous for $x \geq 0$ and differentiable for $x > 0$. It therefore follows from the Mean Value Theorem that for any $x > 0$ there is a number c, $0 < c < x$ such that

$$\frac{f(x) - f(0)}{x - 0} = \frac{\sqrt{1 + x} - 1}{1 + x - 1} = f'(c) = \frac{1}{2\sqrt{1 + c}} < \frac{1}{2}.$$

Hence

$$\frac{\sqrt{1 + x} - 1}{x} < \frac{1}{2}$$

and the desired result follows immediately.

Exercises 12.3

1. Complete the proof of Rolle's Theorem.

2. Prove that if $f : (a, b) \to \mathbb{R}$ is such that $f'(x) \geq 0$ for all x in (a, b), then f is increasing.

3. Prove that if $f : (a, b) \to \mathbb{R}$ is such that $f'(x) \leq 0$ for all x in (a, b), then f is decreasing.

4. Suppose f is continuous on $[a, b]$ and differentiable on (a, b). Suppose further that $|f'(x)| \leq M$ for all $x \in (a, b)$. Prove that $|f(x) - f(y)| \leq M|x - y|$ for all $x, y \in [a, b]$.

5. Suppose $f : \mathbb{R} \to \mathbb{R}$ is everywhere differentiable, $f(0) = 0$, and $|f'(x)| < 1$ for all x. Prove that $|f(x)| < |x|$ for all $x \neq 0$.

6. Suppose $f : \mathbb{R} \to \mathbb{R}$ is everywhere differentiable, $f(a) = 0$ for some a, and $|f'(x)| \leq |f(x)|$ for all x. Prove that $f(x) = 0$ for all x.

7. Prove that $ny^{n-1}(x - y) \leq x^n - y^n \leq nx^{n-1}(x - y)$ if $x \geq y \geq 0$ and $n \geq 1$.

8. Prove that the inequality of Example 12.3.5 also holds for x such that $-1 < x < 0$.

9. Prove that whenever r is a rational number, $0 < r < 1$, then $(1 + h)^r < 1 + rh$ for $h > 0$. What if $h < 0$?

10. Prove that whenever r is a rational number, $1 < r$, then $(1 + h)^r > 1 + rh$ for $h > 0$. What if $h < 0$?

11. Prove that

$$\ln(1 + x) \geq \frac{x}{1 + x}$$

for $x > -1$. (Hint: $(\ln x)' = 1/x$ is a valid assumption.)

12. Prove that $\cos x \geq 1 - (x/2)$ for $x \in [0, \pi/6]$.

13. Let $f : [0, \infty) \to \mathbb{R}$ be a continuous function such that $f(0) = 0$ and $f'(x)$ exists and is increasing for $x > 0$. Prove that the function $f(x)/x$ is monotone increasing for $x > 0$.

14. Suppose $f : (a, b) \to \mathbb{R}$ is differentiable and f'' exists at $t \in (a, b)$. Prove that

$$\lim_{h \to 0} \frac{f(t + h) - 2f(t) + f(t - h)}{h^2} = f''(t).$$

15. Suppose that for some real number c, the function $f : (a, b) \to \mathbb{R}$ satisfies the condition $|f(x) - f(y)| \leq c(x - y)^2$ for all $x, y \in (a, b)$. Prove that $f(x)$ is constant on (a, b).

16. Suppose $f(x) : \mathbb{R} \to \mathbb{R}$ is such that $f(0) = 0$ and $f'(x) \geq f(x)$ for all $x > 0$. Prove that $f(x) \geq 0$ for all $x \geq 0$.

17. Prove Cauchy's Mean Value Theorem: Suppose the functions f and g are continuous on $[a, b]$, differentiable on (a, b), and $g(a) \neq g(b)$. Then there exists a number $c \in (a, b)$ such that

$$\frac{f(b) - f(a)}{g(b) - g(a)} = \frac{f'(c)}{g'(c)}.$$

(Hint: Note that the function $h(t) = [f(b) - f(a)]g(t) - [g(b) - g(a)]f(t)$ has the property that $h(b) = h(a)$ and satisfies the hypotheses of the Mean Value Theorem).

18. Prove L'Hospital's Rule: Suppose the functions f and g are continuous on $[a, b]$ and differentiable on (a, b). If $x^* \in [a, b]$ and

1. $g'(x) \neq 0$ for all $x \in [a, b] - \{x^*\}$,

2. $f(x^*) = g(x^*) = 0$,

3. $\lim\limits_{x \to x^*} \dfrac{f'(x)}{g'(x)} = A$, then $\lim\limits_{x \to x^*} \dfrac{f(x)}{g(x)} = A$.

(Hint: Use Cauchy's Mean Value Theorem.)

19. Use the sequences $\{1 - (1/n)\}$ and $\{-1 + (1/n)\}$ to prove that $f(x) = \sqrt{1 - x^2}$ is not differentiable at $x = \pm 1$.

12.4   INTEGRABILITY

The function $F : D \to \mathbb{R}$ is said to be an *antiderivative* of $f : D \to \mathbb{R}$ if f is the derivative of F. Thus, $x^{r+1}/(r+1)$ is an antiderivative of x^r whenever r is a rational number different from -1, and $-\cos x$ and $\sin x$ are antiderivatives of $\sin x$ and $\cos x$ respectively. A function that has an antiderivative is said to be *integrable*. Accordingly, both x and $\sin x$ are integrable functions. The Dirichlet function of Example 11.2.6 is not integrable, but the proof of this fact falls outside the scope of this text. If F is an antiderivative of f, it is customary to write $F(x) = \int f(x)dx$.

Proposition 12.4.1 *Let F be an antiderivative of $f : [a, b] \to \mathbb{R}$. Then the function G is also an antiderivative of f if and only if there exists a real number C such that $G(x) = F(x) + C$ for all $x \in [a, b]$.*

Proof. Suppose $G(x) = F(x) + C$. Then, by Theorem 12.2.5,

$$G'(x) = F'(x) + C' = f(x) + 0 = f(x)$$

and hence G is an antiderivative of f.

Conversely, suppose G is also an antiderivative of F and set $H(x) = G(x) - F(x)$. Then

$$H'(x) = G'(x) - F'(x) = f(x) - f(x) = 0 \quad \text{for all} \quad x \in [a, b].$$

Hence, by Corollary 12.3.3, $H(x)$ is constant on $[a, b]$, say $H(x) = C$. It now follows that $G(x) = F(x) + C$. □

If F is an antiderivative of f, then we define

$$\int_a^b f(x)dx = F(b) - F(a).$$

It follows from Proposition 12.4.1 that the value of $\int_a^b f(x)dx$ does not depend on the choice of F. In other words, if G is any other antiderivative of f, then

$$\int_a^b f(x)dx = F(b) - F(a) = G(b) - G(a).$$

For later purposes it is important to note that the variable x does not affect the value of $\int_a^b f(x)dx$ either. In other words,

$$\int_a^b f(x)dx = \int_a^b f(t)dt.$$

Proposition 12.4.2 *Let* $f : [a, b] \to \mathbb{R}$ *be an integrable function and* $c \in [a, b]$. *Then the function* $F(x) = \int_c^x f(t)dt$ *is an antiderivative of* f *such that* $F(c) = 0$.

Proof. Let G be any antiderivative of f so that

$$F(x) = \int_c^x f(t)dt = G(x) - G(c).$$

Then $F'(x) = G'(x) - 0 = f(x)$ and $F(c) = G(c) - G(c) = 0$. □

Proposition 12.4.3 *If* $f : [a, b] \to \mathbb{R}$ *is an integrable function such that* $|f(x)| \le M$ *then*

$$\left| \int_a^b f(x)dx \right| \le M(b - a).$$

Proof. Let F be any antiderivative of f. It then follows from the Mean Value Theorem that for some $c \in (a, b)$ we have

$$\left| \int_a^b f(x)dx \right| = |F(b) - F(a)| = |F'(c)(b - a)| = |f(c)| \cdot (b - a) \le M(b - a).$$
□

The proof of the following proposition is relegated to Exercise 1

Proposition 12.4.4 *Suppose f, $g : [a, b] \to \mathbb{R}$ are integrable functions and A and B are real numbers. Then $Af(x) + Bg(x)$ is an integrable function and*

$$\int_a^b [Af(x) + Bg(x)]dx = A \int_a^b f(x)dx + B \int_a^b g(x)dx.$$

In the next chapter repeated use will be made of the following fact whose proof falls outside the scope of this text.

Proposition 12.4.5 *If the function $f : [a, b] \to \mathbb{R}$ is continuous then it is also integrable.*

Exercises 12.4

1. Prove Proposition 12.4.4.

2. Suppose $f, g : [a, b] \to \mathbb{R}$ are integrable functions such that $f(x) \leq g(x)$ for all $x \in [a, b]$. Prove that

$$\int_a^b f(x)dx \leq \int_a^b g(x)dx.$$

(Hint: Define $F(x) = \int_a^x f(t)dt$ and $G(x)$ similarly and use the Mean Value Theorem.)

3. Suppose $f : [a, b] \to \mathbb{R}$ is an integrable function such that $|f(x)|$ is also integrable. Prove that

$$\left| \int_a^b f(x)dx \right| \leq \int_a^b |f(x)|dx.$$

4. Prove that if $f : [a, b] \to \mathbb{R}$ is integrable then there exists $c \in (a, b)$ such that

$$f(c) = \frac{1}{b - a} \int_a^b f(x)dx.$$

5. Prove that if $a < b < c$ and $f : [a, c] \to \mathbb{R}$ is integrable, then

$$\int_a^b f(x)dx + \int_b^c f(x)dx = \int_a^c f(x)dx.$$

6. Suppose the functions $f : [a, b] \to \mathbb{R}$ and $g : [b, c] \to \mathbb{R}$ are integrable and $f(b) = g(b)$. Prove that the function $h : [a, c] \to \mathbb{R}$ defined by $h(x) = f(x)$ for $a \leq x \leq b$ and $h(x) = g(x)$ for $b \leq x \leq c$ is also integrable.

Chapter Summary

In this chapter, a discussion of the evolution of the notion of differentiability is followed by its rigorous definition. A variety of propositions are proved that imply the differentiability of the standard rational and trigonometric functions. The Mean Value Theorem is proved and some of its implications are pointed out. This is followed by a definition of the integral as an antiderivative and a discussion of some of its properties.

13

Uniform Convergence

In this chapter, uniform convergence of infinite series of functions is defined. This concept is used to show that large classes of such infinite series converge to continuous functions and can be both integrated and differentiated term by term.

13.1 UNIFORM AND NON-UNIFORM CONVERGENCE

Sufficient background material has so far been given so that the theoretical properties of Newton's Power Series $\sum c_n x^n$ and Euler's Trigonometric Series $\sum c_n \sin nx$ and $\sum c_n \cos nx$ can now be investigated in depth. Both are particular cases of the general series $\sum g_n(x)$ where $g_n(x)$, $n = 0, 1, 2, \ldots$ are all functions of x over some common domain D. Other examples of such series of functions are

$$\sum_{n=0}^{\infty} \sqrt{1 - x^n} \;=\; 1 + \sqrt{1-x} + \sqrt{1-x^2} + \sqrt{1-x^3} + \cdots$$

and

$$\sum_{n=0}^{\infty} e^{nx} \;=\; 1 + e^x + e^{2x} + e^{3x} + \cdots .$$

It follows from Propositions 9.1.2 and 9.3.2 that the infinite series $\sum \cos nx/n^2$ converges for every value of x and so it is in fact a function, say $f : \mathbb{R} \to \mathbb{R}$.

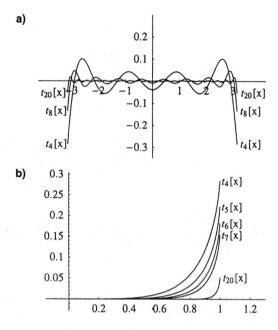

Fig. 13.1   Uniform convergence.

To succumb to temptation and conclude that

$$f'(x) = -\sum \frac{\sin nx}{n}$$

and

$$f''(x) = -\sum \cos nx \tag{1}$$

is to be in error. Whereas the convergence of the series associated with $f'(x)$ is at this point unclear, it has already been shown that the series associated with $f''(x)$ <u>never</u> converges (Example 9.1.4 and Exercise 9.1.3). In other words, statement (1) is at best meaningless. On the other hand, since the geometric series $\sum x^n$ converges in $(-1,1)$ to the value $1/(1-x)$, it is tempting to differentiate term by term twice to arrive at

$$\frac{1}{(1-x)^2} = \sum nx^{n-1}$$

and

$$\frac{1}{(1-x)^3} = \sum n(n-1)x^{n-2}.$$

Both of these series are easily verified to converge in $(-1,1)$ and experimentation with a graphing calculator (Exercises 38, 39) will indicate that these differentiations do lead to correct answers. The fact that these equations are consistent with Newton's Fractional Binomial Theorem (Exercises 3.1.1n, u)

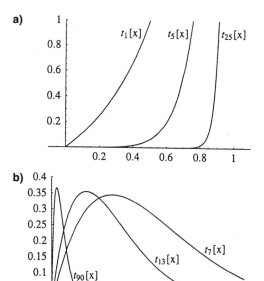

Fig. 13.2   Non-uniform convergence.

also supports the validity of these questionable differentiations. Thus, term-by-term differentiation of infinite series can sometimes lead to correct (and interesting) conclusions, as well as to absurdities. It is this chapter's goal to shed some light on this confusing state of affairs. The crucial concept is that of uniform convergence and it was defined independently in 1841 by Karl Weierstrass, in 1847 by George Stokes (1819–1903), and in 1848 by Phillip Seidel (1821–1896). To explain the notion of uniform convergence of a series of functions $\sum g_n(x)$ it is necessary to focus on the sequence of its tails $\{t_m(x)\}$ where $t_m(x) = \sum_{n=m}^{\infty} g_n(x)$. Figures 13.1 and 13.2 contain the graphs of the tails $t_m(x)$ of four different series of functions:

$$\sum (-1)^{n-1} \frac{\cos nx}{n^2} \qquad -\pi < x < \pi \qquad (13.1a)$$

$$\sum \frac{x^n}{n^2} \qquad 0 \le x \le 1. \qquad (13.1b)$$

$$\sum x^n \qquad 0 \le x < 1 \qquad (13.2a)$$

$$\sum x(1-x)^n (nx + x - 1) \qquad 0 \le x \le .4 \qquad (13.2b)$$

In each of the four diagrams the tails converge to 0 in the sense that at each value $x = a$ of the domain the graphs of the tails get closer and closer to the x-axis. However, in Figure 13.1 the ranges of the tails also shrink to a point whereas the ranges of the tails of Figure 13.2 do not shrink at all. The range of each tail of Figure 13.2a is $[0, \infty)$ whereas the range of the tail $t_m(x)$ of Figure 13.2b is a closed interval $[0, a_m]$ where $\{a_m\}$ is an increasing sequence that converges to a number between .35 and .4. This subtle difference between the manners of convergence displayed in Figures 13.1 and 13.2 is this chapter's key concept.

As was noted in the introductory discussion of Section 9.3, the convergence of any series is tantamount to the convergence of its sequence of tails to zero. A series of functions $\sum g_n(x)$ is said to be *uniformly convergent* in a domain D provided that the ranges of its tails also converge to zero. More formally, provided that for every $\varepsilon > 0$ there is an index m' such that

$$|t_m(x)| < \varepsilon \qquad \text{for all } x \in D \text{ and all } m \geq m'.$$

The term *uniform* can be somewhat misleading. A uniformly convergent series does not necessarily converge at the same rate for all x's. The uniformity in question is the fact that the *same* ε can be used to bound the tails $t_m(x)$ regardless of the specific value of $x \in D$. It may be of interest to note that both Stokes and Seidel referred to a convergence that is not uniform as *infinitely* or *arbitrarily, slow convergence*. In other words, when an infinite series converges non-uniformly, there is always a value x in the domain where an infinite number of the tails have failed to catch up. More precisely, there is an $\varepsilon > 0$, such that for every index m', there is a value x such that $|t_m(x)| \geq \varepsilon$ for some $m > m'$.

The next lemma constitutes the main tool for proving the uniform convergence of a series.

Lemma 13.1.1 (The Weierstrass M-Test) *Suppose $\sum g_n(x)$ is a series of functions and $\sum a_n$ is a series of non-negative numbers such that $|g_n(x)| \leq a_n$ for all indices n and $x \in D$. If the series $\sum a_n$ converges then the series $\sum g_n(x)$ converges uniformly in D.*

Proof. Let $\varepsilon > 0$. Since $\sum a_n$ converges it follows that for sufficiently large m we have $\sum_{n=m}^{\infty} a_n < \varepsilon$. Consequently, by Proposition 9.3.2

$$\left| \sum_{n=m}^{\infty} g_n(x) \right| \leq \sum_{n=m}^{\infty} |g_n(x)| \leq \sum_{n=m}^{\infty} a_n < \varepsilon \qquad \text{for all } x \in D.$$

Thus, the series $\sum g_n(x)$ converges uniformly in D. $\qquad \square$

Example 13.1.2 Consider the series $f(x) = \sum x^2/(n^2 x^2 + 1)$. For $x \neq 0$

$$\frac{x^2}{n^2 x^2 + 1} = \frac{1}{n^2 + \dfrac{1}{x^2}} < \frac{1}{n^2},$$

and for $x = 0$

$$\frac{x^2}{n^2 x^2 + 1} = 0 < \frac{1}{n^2}.$$

It therefore follows from Lemma 13.1.1 and Proposition 9.1.2 that the given series converges uniformly.

Figure 13.2a above indicates that a power series may fail to converge uniformly in its interval of convergence. The next theorem demonstrates that this difficulty is easily alleviated by a slight constriction of the domain.

Theorem 13.1.3 *Suppose the power series* $f(x) = \sum c_n x^n$ *has radius of convergence* ρ. *Then this series converges uniformly on the interval* $[-a, a]$ *whenever* $0 < a < \rho$.

Proof. For any $x \in [-a, a]$,

$$|c_n x^n| = |c_n| \cdot |x|^n \leq |c_n| \cdot a^n = |c_n a^n|.$$

By Theorem 10.1.2 $\sum c_n a^n$ converges absolutely, i.e. $\sum |c_n a^n|$ converges, and hence, by Lemma 13.1.1, $\sum c_n x^n$ converges uniformly in $[-a, a]$. $\qquad\square$

Figure 13.3 displays the uniform convergence of the geometric series on the closed interval $[0, 0.9]$. This should be contrasted with the non-uniform convergence of the same series over the interval $[0, 1]$ which is illustrated in Figure 13.2a.

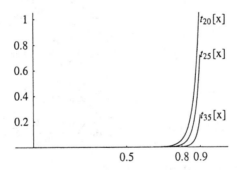

Fig. 13.3   Uniform convergence.

We now turn to an examination of the uniform convergence of trigonometric series. The situation here is considerably more complicated and the result we offer is correspondingly less satisfying. It merely provides a class of series for which uniform convergence can be guaranteed.

Theorem 13.1.4 *Let* $f(x) = \sum c_n \sin nx$ *or* $f(x) = \sum c_n \cos nx$, *be a trigonometric series such that* $c_n = O(n^q)$ *for some* $q < -1$. *Then the series* $f(x)$ *converges uniformly for all values of* x.

Proof. This proof is substantially the same as that of Theorem 10.2.1. Since $\{c_n/n^q\}$ converges, it must be bounded, say

$$|c_n/n^q| \leq M \qquad \text{for all } n.$$

Hence,

$$|c_n \cos nx| \le |c_n| \le Mn^q.$$

The desired result now follows from Lemma 13.1.1 and Proposition 9.1.2. □

Example 13.1.5 Since $1/(n+1)^{3/2} = O\left(n^{-3/2}\right)$ it follows that the series

$$\sum \frac{\sin nx}{(n+1)^{3/2}}$$

converges uniformly for all $x \in \mathbb{R}$.

Exercises 13.1

In each of Exercises 1–20 decide whether the given infinite series converges uniformly in the given domain. Also decide whether the given series converges throughout the given domain. Justify your answer. In the absence of any applicable theorems you may support your answer with suitable computer output.

1. $\displaystyle\sum 2^n x^n$ $D = (-1, 1)$

2. $\displaystyle\sum 2^n x^n$ $D = (-.1, .1)$

3. $\displaystyle\sum 2^n x^n$ $D = (-.5, .5)$

4. $\displaystyle\sum \frac{x^n}{3^n}$ $D = (-1, 1)$

5. $\displaystyle\sum \frac{x^n}{3^n}$ $D = (-3, 3)$

6. $\displaystyle\sum \frac{x^n}{3^n}$ $D = (-2.9, 2.9)$

7. $\displaystyle\sum \frac{x^n}{n^2 3^n}$ $D = (-3, 3)$

8. $\displaystyle\sum \frac{n^2 x^n}{3^n}$ $D = (-3, 3)$

9. $\displaystyle\sum \frac{\sin nx}{n^2}$ $D = (-10, 10)$

10. $\displaystyle\sum \frac{\cos nx}{n(n+1)(n+2)}$ $D = [0, 100]$

11. $\displaystyle\sum e^{nx}$ $D = (-1, 1)$

12. $\displaystyle\sum e^{nx}$ $D = (-3, -2)$

13. $\displaystyle\sum e^{nx}$ $D = (-10, -1)$

14. $\displaystyle\sum \sin^n x$ $D = (-\pi, \pi)$

15. $\displaystyle\sum \frac{\cos e^{nx}}{n^3}$ $D = (-5, 16)$

16. $\displaystyle\sum \frac{\cos(\sin nx)}{n^3}$ $D = (-5, 16)$

17. $\displaystyle\sum \cos\left(\sin \frac{x}{n}\right)$ $D = \mathbb{R}$

18. $\displaystyle\sum \sin\left(\sin \frac{x}{n^2}\right)$ $D = \mathbb{R}$

19. $\sum \dfrac{\sqrt{1-x^{2n}}}{2^n}$ $D = [-1,1]$ 20. $\sum \dfrac{\sqrt{x^{2n}-1}}{2^n}$ $D = [1, \infty)$

21. Prove that if the series $\sum g_n(x)$ converges uniformly over each of the domains D and E separately, then it also converges uniformly over $D \cup E$.

22. Prove that if c is a real number and the series $f(x) = \sum g_n(x)$ converges uniformly over D, then so does the series $\sum c g_n(x)$ converge uniformly to $cf(x)$.

23. Prove that if both the series $g(x) = \sum g_n(x)$ and $h(x) = \sum h_n(x)$ converge uniformly over D, then so does the series $\sum (g_n(x) + h_n(x))$ converge uniformly to $g(x) + h(x)$.

24. Prove that if both the series $g(x) = \sum g_n(x)$ and $h(x) = \sum h_n(x)$ converge uniformly over D, then so does the series $\sum (g_n(x) - h_n(x))$ converge uniformly to $g(x) - h(x)$.

25. Suppose both the series $\sum g_n(x)$ and $\sum h_n(x)$ converge uniformly over D, does $\sum (g_n(x) h_n(x))$ necessarily converge uniformly over D?

26. Show that $\sum x^n(1-x) = 1$ for all $x \in (-1,1]$ and that this convergence is not uniform. (Hint: The tails are easily computed.)

27. Prove that $\sum e^{-(x-n)^2}$ converges uniformly in every interval $[a,b]$. Does it converge uniformly over $\mathbb{R}$?

28. Suppose the series $\sum g_n(x)$ converges absolutely for each value of x. Is the convergence necessarily uniform?

29. Suppose the series $\sum g_n(x)$ converges uniformly in $\mathbb{R}$. Is the convergence necessarily absolute for each value of x?

30. Show that the series $\sum (n^3 + n^4 x^2)^{-1}$ converges uniformly for all x.

31. Prove that the series $\sum x/[n(1 + nx^2)]$ is uniformly convergent for all x. (Hint: Show that the maximum value of the general term is given by $nx^2 = 1$ and use Lemma 13.1.1).

32. Show that the series $\sum (x^n/n^2)(1 + x^{2n})^{-1}$ converges uniformly for all x.

33. Set $g_n(x) = x^n(1 - x^n)$. Show that

a) $\sum g_n(x) = x/(1 - x^2)$ for $|x| < 1$.

b) $\lim_{x \to 1} [\sum g_n(x)] \neq \sum g_n(1)$.

34. Prove that $\sum_{n=m}^{\infty} x^n = x^m/(1 - x)$ for $|x| < 1$ and conclude that each tail of the infinite geometric progression has infinite range.

35. Prove that $\sum x^2/(1 + x^2) \cdot 1/(1 + x^2)^n$ converges uniformly in any closed interval that does not contain 0.

36. Prove that the series $\sum \left(x^{n+1} - x^n \right)$ converges uniformly on the interval $[-a, a]$ provided that $0 < a < 1$.

37. Prove that $\sum (-1)^n / (x^2 + n)$ converges uniformly over $\mathbb{R}$. Where is this convergence absolute?

38. Compare the graphs of $(1-x)^{-2}$ and $\sum_{n=1}^{10} nx^{n-1}$ over the domain $[-.9, 9]$.

39. Compare the graphs of $(1-x)^{-3}$ and $\sum_{n=1}^{10} n(n-1)x^{n-2}$ over the domain $[-.9, 9]$.

40. Let $g_n(x) = x^{n-1}(1 - 2x^n)$ for $n = 1, 2, 3, \ldots$ and $0 \leq x \leq 1$.

   a. Prove that $\sum \int_0^1 g_n(x)dx = 0$ for $n = 1, 2, 3, \ldots$

   b. Prove that $\sum g_n(x) = 1/(1 + x)$ for $0 \leq x \leq 1$.

   c. Prove that $\int_0^1 \left(\sum g_n(x) \right) dx \neq \sum \int_0^1 g_n(x)dx$.

For Exercises 41 and 42, refer to the figures on the following pages.
41. Each of the figures a–k contains the graph of the tail $t_m(x)$ of a series $\sum g_n(x)$ where $g_n : [0, 1] \to \mathbb{R}$. Answer the following questions and justify your answers:

   a. Does the series converge on $[0, 1]$?

   b. Does the series converge uniformly on $[0, 1]$?

   c. Does the series converge on $[0.1, 1]$?

   d. Does the series converge uniformly on $[0.1, 1]$?

42. Each of the figures l–q below contains the graph of the tail $t_m(x)$ of a series $\sum g_n(x)$ where $g_n : [0, 2] \to \mathbb{R}$. Answer the following questions and justify your answers:

   a. Does the series converge on $[0, 2]$?

   b. Does the series converge uniformly on $[0, 2]$?

Each of Exercises 43–47 below specifies a domain D and the general tail $t_m(x)$ of a series $\sum g_n(x)$ that converges on D. Decide whether the series $\sum g_n(x)$ converges uniformly in D.

43. $D = [0, 1]$, $t_m(x) = x^2/m$

44. $D = [0, 1]$, $t_m(x) = x^2/m^2$

45. $D = \mathbb{R}$, $t_m(x) = x^2/m$

46. $D = [0, \infty)$, $t_m(x) = m^2 x/(1 + m^3 x)$

47. $D = [0, \infty)$, $t_m(x) = mx/(1 + m^3 x)$

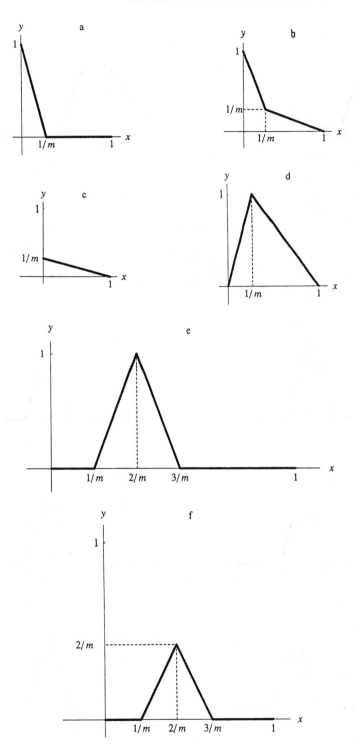

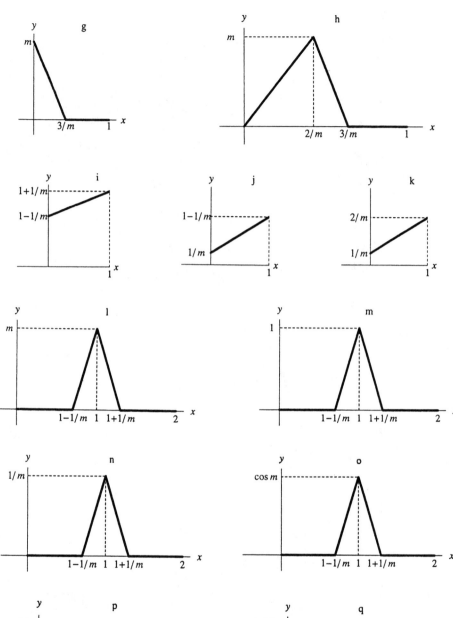
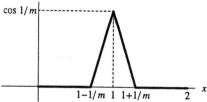
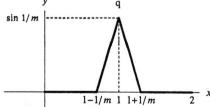

48. Describe the main mathematical achievements of
   a) Philipp Seidel
   b) Georges Stokes

13.2   CONSEQUENCES OF UNIFORM CONVERGENCE

It is now demonstrated that uniformly convergent series have a variety of desirable properties.

Theorem 13.2.1 *Suppose $f(x) = \sum g_n(x)$ is a uniformly convergent series of continuous functions $g_n : D \to \mathbb{R}$. Then $f(x)$ is continuous at each $a \in D$.*

Proof. Set

$$f_m(x) = \sum_{n=0}^{m} g_n(x) \quad \text{and} \quad t_m(x) = \sum_{n=m}^{\infty} g_n(x)$$

so that $f(x) = f_m(x) + t_{m+1}(x)$, $m = 0, 1, 2, 3, \ldots$. Let $\{x_k\} \subset D$, $\{x_k\} \to a$, and let $\varepsilon > 0$. Because of the uniform convergence of the given series, it follows that there is an index m' such that

$$|t_m(x)| < \frac{\varepsilon}{3} \qquad \text{for all } x \in D \text{ and } m \geq m'.$$

Since $f_{m'}(x)$ is a continuous function and $\{x_k\} \to a$, it follows that

$$\lim_{k \to \infty} f_{m'}(x_k) = f_{m'}(a)$$

and so there exists an index k' such that

$$|f_{m'}(x_k) - f_{m'}(a)| < \frac{\varepsilon}{3} \qquad \text{for all } k \geq k'.$$

It now follows that for all $k \geq k'$

$$
\begin{aligned}
|f(x_k) - f(a)| & \\
&= |(f(x_k) - f_{m'}(x_k)) + (f_{m'}(x_k) - f_{m'}(a)) + (f_{m'}(a) - f(a))| \\
&\leq |f(x_k) - f_{m'}(x_k)| + |f_{m'}(x_k) - f_{m'}(a)| + |f_{m'}(a) - f(a)| \\
&= |t_{m'+1}(x_k)| + |f_{m'}(x_k) - f_{m'}(a)| + |t_{m'+1}(a)| \\
&< \frac{\varepsilon}{3} + \frac{\varepsilon}{3} + \frac{\varepsilon}{3} = \varepsilon.
\end{aligned}
$$

Hence $\{f(x_k)\} \to f(a)$ and so $f(x)$ is continuous at each $a \in D$. $\square$

The continuity of most power series and many trigonometric series now follows easily.

Corollary 13.2.2 *If the power series* $f(x) = \sum c_n x^n$ *has radius of convergence* $\rho > 0$, *then it is continuous on* $(-\rho, \rho)$.

Proof. Let a be a number in $(-\rho, \rho)$. Then there exists a number b such that

$$-\rho < -b < a < b < \rho.$$

It follows from Theorem 13.1.3 that $f(x)$ converges uniformly on $[-b, b]$. Consequently, by Theorem 13.2.1, $f(x)$ is continuous at $a \in [-b, b]$.   □

Corollary 13.2.3 *Let* $f(x) = \sum c_n \sin nx$ *or* $f(x) = \sum c_n \cos nx$ *be a trigonometric series such that* $c_n = O(n^q)$ *for some* $q < -1$. *Then the series* $f(x)$ *is continuous in* $\mathbb{R}$.

Proof. This is an immediate consequence of Theorems 13.1.4 and 13.2.1.   □

If the hypothesis that $c_n = O(n^q)$ for some $q < -1$ is <u>not</u> satisfied, the function $f(x)$ may fail to be continuous even when it is well defined (i.e., even when the defining infinite series converges everywhere). Such, for instance is the case for the series

$$f(x) = \sin x - \frac{\sin 2x}{2} + \frac{\sin 3x}{3} - \frac{\sin 4x}{4} + \cdots \qquad (2)$$

which, as is indicated by Figure 13.4, is discontinuous at $x = \pi$. The subtlety of this issue is underscored by the fact that it was in this context that

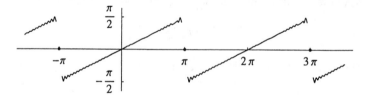

Fig. 13.4   Abel's Discontinuous Limit Function.

Cauchy committed his best known error. In his *Cours d'analyse de l'école polytechnique* Cauchy stated and "proved" the following theorem:

> *When the terms of a series are continuous functions of a variable x in the neighborhood of a particular value where the series converges, the sum of the series is also continuous in that neighborhood.*

His error was pointed out by Niels Abel (1802–1829) in his article on the convergence of the Binomial Series who offered the trigonometric series (Equation 2) as a counterexample to Cauchy's assertion. Note that the series of (2) does converge at $x = \pi$ (to 0), but the sum fails to be continuous there. Ironically, Abel went on to make some errors of his own. The relevance and importance of uniform convergence to this issue was recognized independently by Karl Weierstrass in 1841, George Stokes in 1847, and Phillip Seidel in 1848. However, it was not until Weierstrass discussed it in his 1860 lectures that the significance of uniform convergence was recognized by the mathematical community.

Before leaving this topic we point out that Figure 13.4 supports Euler's claim (made with insufficient proof) that

$$\sin \phi - \frac{\sin 2\phi}{2} + \frac{\sin 3\phi}{3} - \frac{\sin 4\phi}{4} + \cdots = \frac{\phi}{2}$$

at least in the interval $-\pi < \phi < \pi$ (see Proposition 5.3). The validity of this assertion will be established in Section 14.1.

The question of term-by-term integrability, a tool that was used improperly by Euler in the computations displayed in Chapter 5, is now addressed.

Theorem 13.2.4 *Suppose* $f(x) = \sum g_n(x)$ *is a uniformly convergent series of continuous functions* $g_n : [a, b] \to \mathbb{R}$. *Then,* $f(x)$ *is integrable on* $[a, b]$ *and*

$$\int_a^b f(x)dx = \sum \int_a^b g_n(x)dx.$$

Proof. The integrability of $f(x)$ follows from Theorem 13.2.1 and Proposition 12.4.5. For each $m = 0, 1, 2, 3, \ldots$ set

$$f_m(x) = \sum_{n=0}^{m} g_n(x) \quad \text{and} \quad t_m(x) = \sum_{n=m}^{\infty} g_n(x)$$

so that

$$f(x) = f_m(x) + t_{m+1}(x).$$

Let $\varepsilon > 0$. Because of the uniform convergence of the series $\sum g_n(x)$ it follows that for sufficiently large m we have

$$|t_m(x)| < \frac{\varepsilon}{b-a} \qquad \text{for all } x \in [a, b].$$

Consequently, by Section 12.4, for sufficiently large m,

$$\left| \int_a^b f(x)dx - \int_a^b f_m(x)dx \right| = \left| \int_a^b t_{m+1}(x)dx \right| \le \frac{\varepsilon}{b-a} \cdot (b-a) = \varepsilon.$$

It follows that

$$\int_a^b f(x)dx = \lim_{m \to \infty} \int_a^b f_m(x)dx$$

$$= \lim_{m \to \infty} \int_a^b \left(g_0(x) + g_1(x) + g_2(x) + \cdots + g_m(x) \right) dx$$

$$= \lim_{m \to \infty} \left(\int_a^b g_0(x)dx + \int_a^b g_1(x)dx \right.$$

$$\left. + \int_a^b g_2(x)dx + \cdots + \int_a^b g_m(x)dx \right)$$

$$= \sum \int_a^b g_n(x)dx. \qquad \square$$

The conclusion of the above theorem is frequently stated in the form

$$\int_a^b \left(\sum g_n(x) \right) dx = \sum \int_a^b g_n(x)dx.$$

This theorem implies that power series with a positive radius of convergence can be integrated term by term.

Corollary 13.2.5 If $f(x) = \sum c_n x^n$ has radius of convergence $\rho > 0$, then $f(x)$ is integrable and if $\int f(x)dx$ is any antiderivative of $f(x)$, then there is a constant C such that

$$\int f(x)dx = C + \sum \frac{c_n}{n+1} x^{n+1} \qquad \text{for all } x \in (-\rho, \rho).$$

Proof. It follows from Theorems 13.2.4 and 13.1.3 that $f(x)$ is integrable. If $\int f(x)dx$ is any antiderivative of f, then it follows from Propositions 12.4.1-2 that there is a constant of integration C such that

$$\int f(x)dx = C + \int_0^x f(t)dt.$$

Consequently, by Theorem 13.2.4,

$$\int f(x)dx = C + \int_0^x f(t)dt) = C + \int_0^x \left(\sum c_n t^n \right) dt$$

$$= C + \sum \int_0^x c_n t^n dt = C + \sum c_n \frac{x^{n+1}}{n+1} \qquad \square$$

Example 13.2.6 The infinite geometric series yields

$$\frac{1}{1+x^2} = 1 - x^2 + x^4 - x^6 + x^8 - \cdots \qquad \text{for } |x| < 1. \qquad (3)$$

It follows from Corollary 13.2.5 that

$$\int_0^x \frac{1}{1+t^2}dt = C + x - \frac{x^3}{3} + \frac{x^5}{5} - \frac{x^7}{7} + \frac{x^9}{9} - \cdots.$$

The substitution $x = 0$ yields $C = 0$ and hence, if we grant that

$$\int_0^x \frac{1}{1+t^2} dt = \tan^{-1} x$$

then we can conclude that

$$\tan^{-1} x = x - \frac{x^3}{3} + \frac{x^5}{5} - \frac{x^7}{7} + \frac{x^9}{9} - \cdots \qquad \text{for } |x| < 1 \qquad (4)$$

a series whose utility was discussed in Chapter 5.

It is important to note what Corollary 13.2.5 does <u>not</u> say. It makes no assertion whatsoever about the endpoints $x = \pm\rho$. The above example indicates that this issue is subtle and interesting things can happen at these values. The given series of Equation (3) diverges at both $x = \pm 1$ whereas the given function $(1+x^2)^{-1}$ is well defined at both of these values. On the other hand, the integral series of Equation (4) converges at $x = \pm 1$ (Proposition 9.3.8) and the function $\tan^{-1} x$ of (4) is well defined at both ± 1. A consequence of these observations is that we still do <u>not</u> have a verification of the Gregory–Leibniz equation

$$\frac{\pi}{4} = 1 - \frac{1}{3} + \frac{1}{5} - \frac{1}{7} + \cdots$$

obtained by setting $x = 1$ in Equation (4). The verification of the Gregory–Leibniz equation will be given in Corollary 14.1.3. The paradoxical Exercise 14.1.5 shows that the intuitively appealing extension of what is true in $(-a, a)$ to a, even when all the numbers in question are well defined, can lead to error.

The same hypotheses that guarantee the continuity of a trigonometric series also guarantee its term-by-term integrability. The proof is similar to that of Corollary 13.2.5 and is relegated to Exercise 21.

Corollary 13.2.7 *If $f(x) = \sum c_n \cos nx$ (or $\sum c_n \sin nx$) is a trigonometric series and there exists $q < -1$ such that $c_n = O(n^q)$, and if $\int f(x) dx$ is any antiderivative of f, then there exists a constant C such that*

$$\int f(x) dx = C + \sum c_n \frac{\sin nx}{n}$$

or

$$\int f(x) dx = C - \sum c_n \frac{\cos nx}{n}.$$

This corollary justifies some of Euler's integrations in Chapter 5, although by no means all of them. We finally turn to term-by-term differentiation. Unlike the situation for continuity and integrability, a uniformly convergent series of differentiable functions need <u>not</u> be term-by-term differentiable.

Example 13.2.8 It follows from Theorem 13.1.4 that the series

$$\sum (\cos(2n - 1)x)/(2n - 1)^2$$

converges uniformly and its sum is known (see Exercise 14.1.3) to equal $\pi^2/8 - \pi|x|/4$, a function that is clearly not differentiable at $x = 0$. Nevertheless, when this series is differentiated term-by-term, the series

$$\sum -(\sin(2n-1)x)/(2n-1)$$

which does indeed converge to 0 at $x = 0$ is obtained.

This example explains why the statement of Theorem 13.2.9 differs from those of Theorems 13.2.1 and 13.2.4. A function is said to be *continuously differentiable* if its derivative exists and is continuous.

Theorem 13.2.9 *Suppose $f(x) = \sum g_n(x)$ is a convergent series where $g_n(x)$ is continuously differentiable on $[a, b]$ for each $n = 0, 1, 2, \ldots$. If $\sum g'_n(x)$ converges uniformly on $[a, b]$ then $f(x)$ is differentiable on $[a, b]$ and*

$$f'(x) = \sum g'_n(x).$$

Proof. Set $\phi(x) = \sum g'_n(x)$. Since this series is assumed to be uniformly convergent, it follows from Theorem 13.2.4 that $\phi(x)$ can be integrated term by term and so

$$\int_a^x \phi(t)dt = \sum \int_a^x g'_n(t)dt = \sum(g_n(x) - g_n(a)) = f(x) - f(a).$$

It follows that

$$f(x) = f(a) + \int_a^x \phi(t)dt$$

and hence

$$f'(x) = 0 + \left(\int_a^x \phi(t)dt\right)' = \phi(x) = \sum g'_n(x). \qquad \square$$

Theorem 13.2.9 is first used to demonstrate the term-by-term differentiability of power series.

Corollary 13.2.10 *Suppose the power series $f(x) = \sum c_n x^n$ converges on the interval $(-\rho, \rho)$ for some $\rho > 0$. Then $f(x)$ is differentiable on $(-\rho, \rho)$ and*

$$f'(x) = \sum nc_n x^{n-1}.$$

Proof. It suffices show that if a is any real number such that $0 < a < \rho$ then the proposed series $\sum nc_n x^{n-1}$ converges uniformly on $[-a, a]$. Let b be any number such that $a < b < \rho$. Since the series $\sum c_n b^n$ is known to converge absolutely, it suffices to show that

$$\left|nc_n x^{n-1}\right| \le \left|c_n b^n\right|$$

for all sufficiently large n and all $x \in [-a, a]$. Since $|x| \leq a$, it suffices to show that

$$na^{n-1} \leq b^n \quad \text{or} \quad n\left(\frac{a}{b}\right)^{n-1} \leq b \tag{5}$$

for all sufficiently large n. However, the infinite series $\sum n(a/b)^{n-1}$ converges because

$$\lim_{n\to\infty} \frac{(n+1)\left(\frac{a}{b}\right)^n}{n\left(\frac{a}{b}\right)^{n-1}} = \frac{a}{b} \lim_{n\to\infty} \frac{n+1}{n} = \frac{a}{b} < 1.$$

It therefore follows from Proposition 9.1.3.3 that

$$\left\{ n\left(\frac{a}{b}\right)^{n-1} \right\} \to 0$$

and so requirement (5) is clearly satisfied for all sufficiently large n.   □

Example 13.2.11 Since we already know that for $-1 < x < 1$

$$\frac{1}{1+x} = 1 - x + x^2 - x^3 + x^4 - x^5 + \cdots$$

it follows from Proposition 13.2.9 that in the same domain

$$(1+x)^{-2} = -\left(\frac{1}{1+x}\right)' = 1 - 2x + 3x^2 - 4x^3 + 5x^4 - \cdots$$

an expansion that agrees with Newton's Fractional Binomial Theorem (see Exercise 3.1m).

Corollary 13.2.10 validates some aspects of the infinite series solution of differential equations discussed in Section 4.1. Thus, for example, it justifies the assertion that the series $\sum c_n x^n$ whose coefficients were derived in Examples 4.1.1-3 do indeed constitute solutions of the respective proposed differential equations, provided that their radii of convergence are not zero. However, the series of Example 4.1.1 has $\rho = \infty$ because for $n \geq 2$

$$\frac{c_{n+1}}{c_n} = \frac{n!}{(n+1)!} = \frac{1}{n+1} \to 0.$$

That the series of Example 4.1.3 (the Fractional Binomial Series) has $\rho = 1$ was demonstrated in Example 10.1.5. Note that to fully justify the methods of Section 4.1 it is still necessary to prove that in those cases the given equation has a unique solution. This will be done in Section 14.2.

The conditions allowing for term-by-term differentiation of trigonometric series are not as general as those for power series.

Theorem 13.2.12 *Let* $f(x) = \sum c_n \sin nx$ *and* $g(x) = \sum c_n \cos nx$ *be trigonometric series and suppose there exists a real number* $q < -2$ *such that* $c_n = O(n^q)$, *then* $f(x)$ *and* $g(x)$ *are everywhere differentiable and*

$$f'(x) = \sum nc_n \cos nx \quad and \quad g'(x) = -\sum nc_n \sin nx.$$

Proof. Since

$$\{nc_n\} = O(n \cdot n^q) = O\left(n^{q+1}\right)$$

where $q + 1 < -1$, the theorem follows immediately from Theorem 13.1.4 and Theorem 13.2.9. □

Exercises 13.2

In each of Exercises 1–14 try to decide whether the equation is true. If true, cite the relevant proposition. If false, explain why. If you do not have enough information to make a decision say so.

1. $\displaystyle\int \sum \frac{\sin nx}{n} dx = C - \sum \frac{\cos nx}{n^2} \qquad$ for all x.

2. $\displaystyle\left(\sum \frac{\sin nx}{n}\right)' = \sum \cos nx \qquad$ for all x.

3. $\displaystyle\int \sum \frac{\sin nx}{n^2} dx = C - \sum \frac{\cos nx}{n^3} \qquad$ for all x.

4. $\displaystyle\left(\sum \frac{\sin nx}{n^2}\right)' = \sum \frac{\cos nx}{n} \qquad$ for all x.

5. $\displaystyle\int \left(\sum \frac{\sin nx}{n^3}\right) dx = C - \sum \frac{\cos nx}{n^4} \qquad$ for all x.

6. $\displaystyle\left(\sum \frac{\sin nx}{n^3}\right)' = \sum \frac{\cos nx}{n^2} \qquad$ for all x.

7. $\displaystyle\left(\sum \frac{\sin nx}{n^{3/2}}\right)' = \sum \frac{\cos nx}{n^{1/2}} \qquad$ for all x.

8. $\displaystyle\int \left(\sum \frac{\sin nx}{n^{3/2}}\right) dx = C - \sum \frac{\cos nx}{n^{5/2}} \qquad$ for all x.

9. $\displaystyle\left(\sum \frac{\sin nx}{n^{5/2}}\right)' = \sum \frac{\cos nx}{n^{3/2}} \qquad$ for all x.

10. $\displaystyle\int \left(\sum \frac{\sin nx}{n^{5/2}}\right) dx = C - \sum \frac{\cos nx}{n^{7/2}} \qquad$ for all x.

11. $\displaystyle\int \left(\sum x^n\right) dx = C + \sum \frac{x^{n+1}}{n+1} \qquad$ for all x.

12. $\int \left(\sum x^n \right) dx = C + \sum \frac{x^{n+1}}{n+1}$ for all x such that $|x| < 1$.

13. $\left(\sum x^n \right)' = \sum n x^{n-1}$ for $|x| < 1$.

14. $\int \left(\sum e^{nx} \right) dx = C + \sum \frac{e^{nx}}{n}$ for all x.

15. $\left(\sum e^{nx} \right)' = \sum n e^{nx}$ for all x.

16. $\int \left(\sum e^{nx} \right) dx = C + \sum \frac{e^{nx}}{n}$ for all $x < 0$.

17. $\left(\sum e^{nx} \right)' = \sum n e^{nx}$ for all $x < 0$.

18. Suppose $f(x) = \sum a_n x^n$ converges in some interval $(-\rho, \rho)$, $\rho > 0$. Prove that $f(x)$ has derivatives for all orders and $a_n = f^{(n)}(0)/n!$ for each $n = 0, 1, 2, \ldots$.

19. Suppose $\sum a_n x^n$ and $\sum b_n x^n$ are two power series both of which converge in some interval $(-\rho, \rho)$, $\rho > 0$. Prove that if $\sum a_n x^n = \sum b_n x^n$ for all x in $(-\rho, \rho)$, then $a_n = b_n$ for all $n = 0, 1, 2, \ldots$.

20. Suppose $f(x) = \sum c_n \cos nx$ and there exists $q < -1$ such that $c_n = O(n^q)$. Prove that $c_n = 1/\pi \int_{-\pi}^{\pi} f(x) \cos nx \, dx$ for $n = 1, 2, 3 \ldots$. What is the value of c_0?

21. Suppose $f(x) = \sum c_n \sin nx$ and there exists $q < -1$ such that $c_n = O(n^q)$. Prove that $c_n = 1/\pi \int_{-\pi}^{\pi} f(x) \sin nx \, dx$ for $n = 1, 2, 3, \ldots$.

22. Prove that the assumption that $f(x) = \sum g_n(x)$ is a convergent series in Theorem 13.2.9 can be replaced by the weaker condition that $\sum g_n(a)$ converges.

23. Prove Newton's Binomial Theorem when r is a negative integer. (Hint: Use mathematical induction.)

24. Prove Corollary 13.2.7.

25. Suppose $\sum g_n(x)$ converges uniformly on $[a, b]$. If each g_n is continuous, prove that the series $\sum \int_a^x g_n(t) \, dt$ also converges uniformly on $[a, b]$.

26. Suppose $\sum g_n(x)$ is a series such that $\sum g_n'(x)$ converges uniformly. Must $\sum g_n(x)$ also converge uniformly?

27. Explain why there is no power series $\sum c_n x^n$ such that $|x| = \sum c_n x^n$ for all $x \in (-\rho, \rho)$ where ρ is some positive number.

28.

    a) Prove that $\sum nx^n = x/(1-x)^2$ for $|x| < 1$.

    b) Evaluate $\sum n/2^n$.

    c) Evaluate $\sum (-1)^n n/3^n$.

29.

    a) Evaluate $\sum n^2 x^n$ as a rational function of x for $|x| < 1$.

    b) Evaluate $\sum n^2/2^n$.

30. Prove that the series $\sum \left(x^{n+1} - x^n\right)$ does not converge uniformly on $[0, 1]$.

31. Prove that the series $\sum \left(x^{n+1} - x^n\right)$ does not converge uniformly on $(0, 1)$.

32. Prove that the series $\sum (\cos nx)/2^n$ converges to a continuous and differentiable function.

33. Prove that the series $\sum_{n=0}^{\infty} x^2/(1 + x^2) \cdot 1/(1 + x^2)^n$ does not converge uniformly on $[-1, 1]$.

34. Prove that the series $\sum (x - n)^{-2}$ converges to a function that is differentiable in its domain. Describe this domain.

35. Describe the main mathematical achievements of Niels Abel.

Chapter Summary

This chapter begins with a definition and a discussion of the evolution of uniform convergence of an infinite series of functions over a domain. Sufficient conditions are described which guarantee the uniform convergence of power series over closed subintervals of their interval of convergence and of some trigonometric series. It is then demonstrated that the uniform convergence of these infinite series is sufficient to imply the continuity, integrability, and differentiability of power series and many trigonometric series.

14

The Vindication

Rigorous proofs are offered for several of the theorems stated in the first part of this book.

14.1 TRIGONOMETRIC SERIES

We return to Euler's results of Chapter 5 and give them a logical grounding. First a finite analog of Equation (5.4) is proved.

Lemma 14.1.1 *If n is any non-negative integer, then*

$$\cos\phi - \cos 2\phi + \cos 3\phi - \cdots - (-1)^n \cos n\phi = \frac{1}{2} - (-1)^n \frac{\cos\left(n + \frac{1}{2}\right)\phi}{2\cos\frac{\phi}{2}}.$$

Proof. By induction on n. If $n = 0$, then both sides of the equation reduce to 0 and so the induction is anchored at $n = 0$. Assume the lemma to hold for a certain specific value of n. Then the identity $2\cos\alpha\cos\beta = \cos(\alpha + \beta) + \cos(\alpha - \beta)$ yields

$$\cos\phi - \cos 2\phi + \cos 3\phi - \cdots - (-1)^n \cos n\phi - (-1)^{n+1}\cos(n+1)\phi$$

$$= \frac{1}{2} - (-1)^n \left(\frac{\cos\left(n + \frac{1}{2}\right)\phi}{2\cos\frac{\phi}{2}} - \cos(n+1)\phi \right)$$

$$= \frac{1}{2} - (-1)^n \frac{\cos\left(n + \frac{1}{2}\right)\phi - 2\cos(n+1)\phi\cos\frac{\phi}{2}}{2\cos\frac{\phi}{2}}$$

$$= \frac{1}{2} - (-1)^n \frac{\cos\left(n + \frac{1}{2}\right)\phi - \cos\left(n + \frac{3}{2}\right)\phi - \cos\left(n + \frac{1}{2}\right)\phi}{2\cos\frac{\phi}{2}}$$

$$= \frac{1}{2} - (-1)^{n+1} \frac{\cos\left(n + 1 + \frac{1}{2}\right)\phi}{2\cos\frac{\phi}{2}}.$$

Hence, the induction is complete. □

The easy inductive proof of the above lemma has the disadvantage of obscuring the identity's origin. An alternate proof can be based on the substitution of $z = -(\cos\phi + i\sin\phi)$ in the finite geometric progression

$$1 + z + z^2 + z^3 + \cdots + z^n = \frac{1 - z^{n+1}}{1 - z}.$$

and an application of De Moivre's Theorem. While this looks very much like a repetition of Euler's trick, it only involves finite sums and is therefore legitimate.

The next identity is of a type that has many useful applications both in pure and applied mathematics.

Proposition 14.1.2 *If $-\pi < -a \le x \le a < \pi$, then*

$$\frac{x}{2} = \sin x - \frac{\sin 2x}{2} + \frac{\sin 3x}{3} - \frac{\sin 4x}{4} + \cdots$$

and the convergence is uniform on $[-a, a]$.

Proof. For every positive integer n Lemma 14.1.1 yields

$$\sin x - \frac{\sin 2x}{2} + \frac{\sin 3x}{3} - \frac{\sin 4x}{4} + \cdots - (-1)^n \frac{\sin nx}{n}$$

$$= \int_0^x \cos\phi d\phi - \int_0^x \cos 2\phi d\phi + \int_0^x \cos 3\phi d\phi - \cdots - (-1)^n \int_0^x \cos n\phi d\phi$$

$$= \int_0^x \left(\frac{1}{2} - (-1)^n \frac{\cos\left(n + \frac{1}{2}\right)\phi}{2\cos\frac{\phi}{2}}\right) d\phi$$

$$= \frac{x}{2} - (-1)^n \int_0^x \frac{\cos\left(n + \frac{1}{2}\right)\phi}{2\cos\frac{\phi}{2}} d\phi.$$

It is now clear that our goal must be to show that

$$\lim_{n\to\infty}\int_0^x \frac{\cos\left(n+\dfrac{1}{2}\right)\phi}{2\cos\dfrac{\phi}{2}}\,d\phi = 0$$

independently of the value of x (as long as x is in $[-a, a]$). This, however, is a surprisingly easy task. Since $-\pi < -a \le -x \le \phi \le x \le a < \pi$ and $\cos(-\alpha) = \cos\alpha$, it follows that $\cos\phi/2 \ge \cos a/2 > 0$ and hence

$$\left|\int_0^x \frac{\cos\left(n+\dfrac{1}{2}\right)\phi}{2\cos\dfrac{\phi}{2}}\,d\phi\right| \le \frac{1}{2\cos\dfrac{a}{2}}\left|\int_0^x \cos\left(n+\frac{1}{2}\right)\phi\,d\phi\right|$$

$$= \frac{1}{2\cos\dfrac{a}{2}}\cdot\frac{\left|\sin\left(n+\dfrac{1}{2}\right)x\right|}{\left(n+\dfrac{1}{2}\right)} \le \frac{1}{(2n+1)\cos\dfrac{a}{2}}.$$

Thus, as n grows indefinitely, the integral converges to 0 uniformly in $[-a, a]$. $\qquad\square$

The substitution of $x = \pi/2$ into the identity of Proposition 14.1.2 yields the Gregory–Leibniz equation of Chapter 4.

Corollary 14.1.3 $\pi/4 = 1 - 1/3 + 1/5 - 1/7 + \cdots$.

Many more interesting expansions of this type can be derived. One more is obtained below and the derivation of others is outlined in Exercises 3, 4.

Corollary 14.1.4 *If $-\pi < x < \pi$, then*

$$\frac{x^2}{4} = \frac{\pi^2}{12} - \cos x + \frac{\cos 2x}{2^2} - \frac{\cos 3x}{3^2} + \frac{\cos 4x}{4^2} - \cdots .$$

Proof. Let $x \in [-a, a]$ where $0 < a < \pi$. Then, by the previous proposition, Theorem 13.2.4, and Proposition 9.1.3.2,

$$\frac{x^2}{4} = \int_0^x \frac{t}{2}\,dt$$

$$= \int_0^x \sin t\,dt - \int_0^x \frac{\sin 2t}{2}\,dt + \int_0^x \frac{\sin 3t}{3}\,dt - \int_0^x \frac{\sin 4t}{4}\,dt + \cdots$$

$$= -\cos t\Big|_0^x + \frac{\cos 2t}{2^2}\Big|_0^x - \frac{\cos 3t}{3^2}\Big|_0^x + \frac{\cos 4t}{4^2}\Big|_0^x - \cdots$$

$$= -(\cos x - 1) + \frac{\cos 2x - 1}{2^2} - \frac{\cos 3x - 1}{3^2} + \frac{\cos 4x - 1}{4^2} - \cdots$$

$$= (1 - \frac{1}{2^2} + \frac{1}{3^2} - \frac{1}{4^2} + \cdots)$$

$$- \left(\cos x - \frac{\cos 2x}{2^2} + \frac{\cos 3x}{3^2} - \frac{\cos 4x}{4^2} - \cdots \right) \qquad (3)$$

The rest of the proof is the same as that of Proposition 5.1. Proposition 9.3.8 guarantees the convergence of $1 - 1/2^2 + 1/3^2 - 1/4^2 + \cdots$, say to C_2. If we substitute $x = \pi/2$ into Equation (3) we get

$$\frac{\pi^2}{16} = \frac{(\pi/2)^2}{4} = C_2 - \left(\cos \frac{\pi}{2} - \frac{\cos \pi}{2^2} + \frac{\cos \frac{3\pi}{2}}{3^2} - \frac{\cos 2\pi}{4^2} + \cdots \right)$$

$$= C_2 - \left(0 - \frac{-1}{2^2} + \frac{0}{3^2} - \frac{1}{4^2} + \frac{0}{5^2} - \frac{-1}{6^2} + \frac{0}{7^2} - \frac{1}{8^2} + \cdots \right)$$

$$= C_2 - \left(\frac{1}{2^2} - \frac{1}{4^2} + \frac{1}{6^2} - \frac{1}{8^2} + \cdots \right)$$

$$= C_2 - \frac{1}{4} \left(1 - \frac{1}{2^2} + \frac{1}{3^2} - \frac{1}{4^2} + \cdots \right)$$

$$= C_2 - \frac{1}{4} C_2 = \frac{3}{4} C_2.$$

Hence,

$$C_2 = \frac{4}{3} \cdot \frac{\pi^2}{16} = \frac{\pi^2}{12}$$

and the desired conclusion follows from Equation (3).   □

Corollary 14.1.5

1. $\dfrac{\pi^2}{12} = 1 - \dfrac{1}{2^2} + \dfrac{1}{3^2} - \dfrac{1}{4^2} + \cdots$.

2. $\dfrac{\pi^2}{6} = 1 + \dfrac{1}{2^2} + \dfrac{1}{3^2} + \dfrac{1}{4^2} + \cdots$.

Proof. Equation 1 follows from the substitution $x = 0$ into the identity of Corollary 14.1.4. The proof of Equation 2 is identical with that offered in Proposition 5.2.

Exercises 14.1

1. Prove rigorously that for $-\pi < x < \pi$

a) $\dfrac{x^3}{12} - \dfrac{\pi^2 x}{12} = -\sin x + \dfrac{\sin 2x}{2^3} - \dfrac{\sin 3x}{3^3} + \dfrac{\sin 4x}{4^3} - \cdots$

b) $\dfrac{x^4}{48} - \dfrac{\pi^2 x^2}{24} + \dfrac{\pi^4}{720} = \cos x - \dfrac{\cos 2x}{2^4} + \dfrac{\cos 3x}{3^4} - \dfrac{\cos 4x}{4^4} + \cdots$

2. Prove that

$$\cos\phi + \cos 2\phi + \cos 3\phi + \cdots + \cos n\phi = -\frac{1}{2} + \frac{\sin\left(n + \dfrac{1}{2}\right)\phi}{2\sin\dfrac{\phi}{2}}.$$

(Hint: Replace ϕ with $\phi + \pi$ in the statement of Lemma 14.1.1)

3. a) Prove that for odd n

$$\cos\phi + \cos 3\phi + \cos 5\phi + \cdots + \cos n\phi = \frac{\sin(n+1)\phi}{2\sin\phi}.$$

(Hint: Exercise 2 can be used.)

b) Prove that for any numbers $0 < a, b < \pi$,

$$\lim_{n\to\infty} \int_a^b \frac{\sin(n+1)\phi}{2\sin\phi} d\phi = 0.$$

c) Prove that the series

$$\sin x + \frac{\sin 3x}{3} + \frac{\sin 5x}{5} + \frac{\sin 7x}{7} + \cdots$$

converges for all x such that $0 < |x| < \pi$, and that it has a constant value on each of the intervals $(0, \pi)$ and $(-\pi, 0)$ (not on both, just on each).

d) Prove that the constant values of part c are $\pi/4$ and $-\pi/4$, respectively.

e) Prove that if $-\pi < x < \pi$ then

$$|x| = \frac{\pi}{2} - \frac{4}{\pi}\left(\cos x + \frac{\cos 3x}{3^2} + \frac{\cos 5x}{5^2} + \frac{\cos 7x}{7^2} + \cdots\right).$$

4. a) Prove that for every non-negative integer n

$$\sin\phi - \sin 3\phi + \sin 5\phi - \sin 7\phi + \cdots + (-1)^n \sin(2n+1)\phi$$
$$= (-1)^n \frac{\sin(2n+2)\phi}{2\cos\phi}.$$

b) Prove that if $-\pi/2 < a < b < \pi/2$, then $\displaystyle\lim_{n\to\infty} \int_a^b \frac{\sin(2n+2)\phi}{2\cos\phi} d\phi = 0$.

c) Prove that the series

$$\cos x - \frac{\cos 3x}{3} + \frac{\cos 5x}{5} - \frac{\cos 7x}{7} + \cdots$$

converges for each x, $-\pi < x < \pi$ and that it has a constant value on each of the intervals $(-\pi, -\pi/2)$, $(-\pi/2, \pi/2)$, $(\pi/2, \pi)$.

d) Compute the value of

$$\cos x - \frac{\cos 3x}{3} + \frac{\cos 5x}{5} - \frac{\cos 7x}{7} + \cdots$$

for each $x \in (-\pi, \pi)$.

e) For each $x \in (-\pi, \pi)$ compute the value of

$$\sin x - \frac{\sin 3x}{3^2} + \frac{\sin 5x}{5^2} - \frac{\sin 7x}{7^2} + \cdots .$$

5. Resolve the following paradox:

$$\frac{\pi}{2} = \sin \pi - \frac{\sin 2\pi}{2} + \frac{\sin 3\pi}{3} - \frac{\sin 4\pi}{4} + \cdots$$
$$= 0 - \frac{0}{2} + \frac{0}{3} - \frac{0}{4} + \cdots = 0.$$

14.2 POWER SERIES

The results demonstrated in Chapter 13 are now used to justify some well known power series expansions, including Newton's Fractional Binomial Theorem. In each case both the function and its power series are shown to satisfy the same differential equations. Lemmas that guarantee the uniqueness of the solutions of these differential equations conclude that in each case the function and its power series expansion are necessarily equal.

Lemma 14.2.1 *Suppose $f : [a, b] \to \mathbb{R}$ is differentiable, $f(a) = 0$, and $|f'(x)| \le |f(x)|$. Then $f(x) = 0$ for all $x \in [a, b]$.*

Proof. We first assume that $b < a + 1$. By the Maximum Principle, there exists a number $x_M \in [a, b]$ such that $|f(x_M)|$ is the maximum of $|f(x)|$ on $[a, b]$. If $x_M = a$, then the fact that $f(a) = 0$ implies the desired conclusion. If $x_M > a$, then the Mean Value Theorem guarantees the existence of a number c, $a < c < x_M$, such that

$$\frac{f(x_M) - f(a)}{x_M - a} = f'(c).$$

Hence,

$$|f(x_M)| = |f(x_M) - f(a)| = |f'(c)| \cdot |x_M - a|$$
$$\le |f(c)| \cdot |x_M - a| \le |f(x_M)| \cdot |x_M - a|.$$

Since $|x_M - a| \le |b - a| < 1$ it now follows that $f(x_M) = 0$ and hence $f(x) = 0$ for all $x \in [a, b]$. Since f is continuous, the lemma is also true when $b = a + 1$.

If b is arbitrary, then a combination of the above argument with a straightforward induction allows the conclusion that $f(x) = 0$ on each of the successive intervals $[a, a + 1]$, $[a + 1, a + 2], \cdots , [a + k - 1, a + k]$ where k is the largest integer such that $a + k \in [a, b]$. One more application of the first part of this proof to the interval $[a + k, b]$ then concludes the proof.   □

Make the assumption that the exponential function e^x equals its own derivative. A more detailed treatment of this function will be found in Exercise 4.

Corollary 14.2.2 $e^x = 1 + x + x^2/2! + x^3/3! + \cdots$ *for all $x \in \mathbb{R}$.*

Proof. If $p(x) = 1 + x + x^2/2! + x^3/3! + \cdots$ then it follows from Corollary 13.2.10 that

$$p'(x) = 0 + 1 + 2 \cdot \frac{x}{2!} + 3 \cdot \frac{x^2}{3!} + 4 \cdot \frac{x^3}{4!} + \cdots$$

$$= 1 + x + \frac{x^2}{2!} + \frac{x^3}{3!} + \cdots = p(x).$$

Now set $f(x) = e^x - p(x)$ and observe that $f(0) = 0$ and

$$|f'(x)| = |e^x - p(x)| = |f(x)|$$

so that the function $f(x)$ satisfies the hypotheses of Lemma 14.2.1 on every interval $[0, b]$. It follows that $f(x) = 0$ for all $x \ge 0$ and hence

$$e^x = 1 + x + \frac{x^2}{2!} + \frac{x^3}{3!} + \cdots \quad \text{for all} \quad x \ge 0.$$

The case $x < 0$ is relegated to Exercise 8.   □

The series expansion of e^x yields, of course, good estimates of e. Surprisingly, it also provides the basis for a proof of the irrationality of e.

Corollary 14.2.3 *The number e is irrational.*

Proof. Suppose, to the contrary, that $e = s/t$ where s and t are positive integers. If m is any integer greater than t then

$$1 + 1 + \frac{1}{2!} + \frac{1}{3!} + \cdots + \frac{1}{(m-1)!} < e^1 = \frac{s}{t}$$

$$= 1 + 1 + \frac{1}{2!} + \frac{1}{3!} + \cdots + \frac{1}{(m-1)!} + \frac{1}{m!} + \frac{1}{(m+1)!} + \frac{1}{(m+2)!} + \cdots$$

$$< 1 + 1 + \frac{1}{2!} + \frac{1}{3!} + \cdots + \frac{1}{(m-1)!} + \frac{1}{m!} \left(1 + \frac{1}{m} + \frac{1}{m^2} + \cdots \right)$$

$$= 1 + 1 + \frac{1}{2!} + \frac{1}{3!} + \cdots + \frac{1}{(m-1)!} + \frac{1}{m!} \cdot \frac{1}{1 - \frac{1}{m}}$$

$$= 1 + 1 + \frac{1}{2!} + \frac{1}{3!} + \cdots + \frac{1}{(m-1)!} + \frac{1}{(m-1)!(m-1)}.$$

When the above inequalities are multiplied by $(m-1)!$ we get

$$(m-1)! + (m-1)! + \frac{(m-1)!}{2!} + \frac{(m-1)!}{3!} + \cdots + \frac{(m-1)!}{(m-1)!}$$

$$< \frac{(m-1)!s}{t} < (m-1)! + (m-1)! + \frac{(m-1)!}{2!}$$

$$+ \frac{(m-1)!}{3!} + \cdots + \frac{(m-1)!}{(m-1)!} + \frac{1}{m-1}.$$

Since each of the summands of

$$\mu = (m-1)! + (m-1)! + \frac{(m-1)!}{2!} + \frac{(m-1)!}{3!} + \cdots + \frac{(m-1)!}{(m-1)!}$$

is an integer, so is μ an integer. Similarly, since $m > t$ it follows that $(m-1)!s/t$ is also an integer, say ν. Thus, we have integers μ, ν such that

$$\mu < \nu < \mu + \frac{1}{m-1}$$

which is absurd since $1/(m-1)) < 1$. Hence e is irrational.   □

The proofs of 14.2.4 and 14.2.5 below are similar to those above and are relegated to the Exercises 5, 6.

Lemma 14.2.4 *Suppose $f : [a, b] \to \mathbb{R}$ is twice differentiable, $f(a) = f'(a) = 0$, and $|f''(x)| \le |f(x)|$. Then $f(x) = 0$ for all $x \in [a, b]$.*

Corollary 14.2.5 *For all $x \in \mathbb{R}$*

1. $\sin x = x - \dfrac{x^3}{3!} + \dfrac{x^5}{5!} - \dfrac{x^7}{7!} + \cdots$

2. $\cos x = 1 - \dfrac{x^2}{2!} + \dfrac{x^4}{4!} - \dfrac{x^6}{6!} + \cdots .$

The proof of the Fractional Binomial Theorem requires a more general version of Lemma 14.2.1.

Lemma 14.2.6 *Suppose $f : [a, b] \to \mathbb{R}$ is differentiable, $f(a) = 0$, and $|f'(x)| \le A|f(x)|$ for some number A. Then $f(x) = 0$ for all $x \in [a, b]$.*

Proof. Clearly $A \ge 0$. If $A = 0$ then $f'(x) = 0$ for each $x \in [a, b]$ and the conclusion follows from Corollary 12.3.3. Otherwise, set

$$g(x) = f\left(\frac{x}{A}\right) \qquad x \in [aA, bA].$$

Since

$$|g'(x)| = \left| \frac{1}{A} \cdot f'\left(\frac{x}{A}\right) \right| \le \frac{1}{A} \cdot \left| f'(\frac{x}{A}) \right|$$

$$\le \frac{1}{A} \cdot A \cdot \left| f\left(\frac{x}{A}\right) \right| = |g(x)|,$$

it follows that the function $g(x)$ satisfies the hypotheses of Lemma 14.2.1. Consequently $g(x)$ vanishes on $[aA, bA]$ and so $f(x)$ vanishes on $[a, b]$.   □

Corollary 14.2.7 (The Fractional Binomial Theorem)
If r is any rational number, then

$$(1 + x)^r = 1 + \binom{r}{1}x + \binom{r}{2}x^2 + \binom{r}{3}x^3 + \cdots \qquad |x| < 1.$$

Proof. For $|x| < 1$ set $p(x) = 1 + \binom{r}{1}x + \binom{r}{2}x^2 + \binom{r}{3}x^3 + \cdots$ (see Example 10.1.5). It was demonstrated in Example 4.1.3 that

$$(1 + x)p'(x) = r \cdot p(x).$$

Consequently, if we set $f(x) = (1 + x)^r - p(x)$ in the interval $[a, b]$ where $-1 < a < b < 1$, then $f(0) = 1 - 1 = 0$ and

$$|f'(x)| = \left| r(1 + x)^{r-1}_, - \frac{r \cdot p(x)}{1 + x} \right| = \frac{|r|}{1 + x} \cdot |(1 + x)^r_, - p(x)|$$

$$= \frac{|r|}{1 + x} \cdot |f(x)| < \frac{|r|}{1 + a} \cdot |f(x)|.$$

It therefore follows from Lemma 14.2.6 that $f(x)$ vanishes in every interval of the form $[a, b]$, $-1 < a < b < 1$. Thus, $(1 + x)^r = p(x)$ for $|x| < 1$.   □

Exercises 14.2

1. Prove that $\sin 1$ is irrational.

2. Prove that $\cos 1$ is irrational.

3. Prove that $\sum \frac{a_n}{n!}$ is irrational whenever $\{a_n\}$ is a bounded sequence of positive integers.

4. An alternative approach to the function e^x is the following. Define a function $\text{Exp}(x) = 1 + \frac{x}{1!} + \frac{x^2}{2!} + \frac{x^3}{3!} + \cdots$ and set $e = \text{Exp}(1)$.

   a) Prove that $\text{Exp}(x)$ converges for all x.

   b) Prove that $\text{Exp}'(x) = \text{Exp}(x)$.

c) Use Proposition 14.2.6 to prove that for any real numbers a and x $\text{Exp}(a) \cdot \text{Exp}(x) = \text{Exp}(a + x)$.

d) Prove that for every rational number x, $\text{Exp}(x) = e^x$.

e) Assuming that for every real number b the function $b^x : \mathbb{R} \to \mathbb{R}$ has been defined as a continuous function, prove that $\text{Exp}(x) = e^x$ for all $x \in \mathbb{R}$.

5. Prove Lemma 14.2.4.

6. Prove Corollary 14.2.5.

7. Explain why the following differential equations have at most one solution in the specified domain.

a) $y' = x - y$, $y(0) = 1$, $D = \mathbb{R}$

b) $y' = 1 + x + y$, $y(0) = 1$, $D = \mathbb{R}$

c) $y' = 2 - x + xy$, $y(0) = 1$, $D = \mathbb{R}$

d) $1 + 2x - 2y + 3y' = xy$, $y(0) = 1$, $D = \mathbb{R}$

e) $3 - x^2 + 2xy = y' + \dfrac{1}{1 - x}$, $y(0) = 1$, $D = (-1, 1)$

f) $xy' = 1 + y$, $y(1) = 1$, $D = (0, \infty)$

g) $x^2 y' = 1 - y$, $y(1) = 1$, $D = (0, \infty)$

8. Complete the proof of Corollary 14.2.2.

Chapter Summary

Uniform convergence is used to justify those trigonometric series of Euler's that are in fact valid. The Maximum Principle and the Mean Value Theorem are used to prove two propositions that guarantee the uniqueness of solutions of some differential equations. This uniqueness, together with the above justified differentiabilty of power series, demonstrates the validity of Newton's Fractional Binomial Theorem and of the power series expansions of e^x, $\sin x$ and $\cos x$.

Appendix A

Excerpts from
"Quadrature of the Parabola"
by Archimedes

Archimedes to Dositheus greeting.

When I heard that Conon, who was my friend in his lifetime, was dead, but that you were acquainted with Conon and withal versed in geometry, while I grieved for the loss not only of a friend but of an admirable mathematician, I set myself the task of communicating to you, as I had intended to send to Conon, a certain geometrical theorem which had not been investigated before but has now been investigated by me, and exhibited by means of geometry. Now some of the earlier geometers tried to prove it possible to find a rectilineal area equal to a given circle and a given segment of a circle; and after that they endeavoured to square the area bounded by the section of the whole cone and a straight line, assuming lemmas not easily conceded, so that it was recognised by most people that the problem was not solved. But 1 am not aware that any one of my predecessors has attempted to square the segment bounded by a

Reprinted, with permission, from *The Works of Archimedes*, T. L. Heath, ed., Cambridge University Press, Cambridge, United Kingdom, 1897. Excerpts are reproduced word for word; ends of excerpts are indicated by three asterisks.

straight line and a section of a right-angled cone [a parabola], of which problem I have now discovered the solution. For it is here shown that every segment bounded by a straight line and a section of a right-angled cone [a parabola] is four-thirds of the triangle which has the same base and equal height with the segment, and for the demonstration of this property the following lemma is assumed: that the excess by which the greater of (two) unequal areas exceeds the less can, by being added to itself, be made to exceed any given finite area. The earlier geometers have also used this lemma; for it is by the use of this same lemma that they have shown that circles are to one another in the duplicate ratio of their diameters, and further that every pyramid is one third part of the prism which has the same base with the pyramid and equal height; also, that every cone is one third part of the cylinder having the same base as the cone and equal height they proved by assuming a certain lemma similar to that aforesaid. And, in the result, each of the aforesaid theorems has been accepted no less than those proved without the lemma. As therefore my work now published has satisfied the same test as the propositions referred to, I have written out the proof and send it to you, first as investigated by means of mechanics, and afterwards too as demonstrated by geometry. Prefixed are, also, the elementary propositions in conics which are of service in the proof. Farewell."

Proposition 1. *If from a point on a parabola a straight line be drawn which is either itself the axis or parallel to the axis, as PV, and if QQ' be a chord parallel to the tangent to the parabola at P and meeting PV in V, then*

$$QV = VQ'.$$

Conversely, if QV = VQ', the chord QQ' will be parallel to the tangent at P.

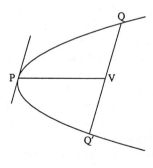

* * *

Proposition 3. *If from a point on a parabola a straight line be drawn which is either itself the axis or parallel to the axis, as PV, and if from two other points Q, Q' on the parabola straight lines be drawn parallel to the tangent at*

P and meeting PV in V, V' respecively, then

$$PV : PV' = QV^2 : Q'V'^2.$$

"And these propositions are proved in the elements of conics."

*   *   *

Definition. "In segments bounded by a straight line and any curve I call the straight line the **base**, and the **height** the greatest perpendicular drawn from the curve to the base of the segment, and the **vertex** the point from which the greatest perpendicular is drawn."

Proposition 18. *If Qq be the base of a segment of a parabola, and V the middle point of Qq, and if the diameter through V meet the curve in P, then P is the vertex of the segment.*

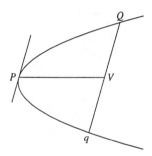

For *Qq* is parallel to the tangent at *P* [Prop. 1]. Therefore, of all the perpendiculars which can be drawn from points on the segment to the base *Qq*, that from *P* is the greatest. Hence, by definition, *P* is the vertex of the segment.

Proposition 19. *If Qq be a chord of a parabola bisected in V by the diameter PV, and if RM be a diameter bisecting QV in M, and RW be the ordinate from R to PV, then*

$$PV = \frac{4}{3}RM.$$

For, by property of the parabola,

$$PV : PW = QV^2 : RW^2$$
$$= 4RW^2 : RW^2$$

so that

$$PV = 4PW$$

whence

$$PV = \frac{4}{3}RM.$$

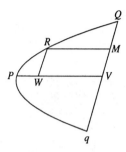

Proposition 20. *If Qq be the base, and P the vertex, of a parabolic segment, then the triangle PQq is greater than half the segment PQq.* For the chord Qq

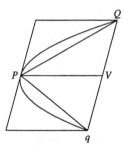

is parallel to the tangent at P, and the triangle PQq is half the parallelogram formed by Qq, the tangent at P, and the diameters through Q, q.

Therefore the triangle PQq is greater than half the segment.

Corollary. It follows that *it is possible to inscribe in the segment a polygon such that the segments left over are together less than any assigned area.*

Proposition 21. *If Qq be the base, and P the vertex, of any parabolic segment, and if R be the vertex of the segment cut off by PQ, then*

$$\triangle PQq = 8 \triangle PRQ.$$

The diameter through R will bisect the chord PQ, and therefore also QV, where PV is the diameter bisecting Qq. Let the diameter through R bisect PQ in Y and QV in M. Join PM.

By Prop. 19

$$PV = \frac{4}{3}RM.$$

Also

$$PV = 2YM.$$

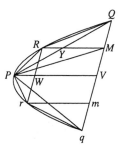

Therefore
$$YM = 2RY,$$
and
$$\triangle PQM = 2 \triangle PRQ.$$
Hence
$$\triangle PQV = 4 \triangle PRQ,$$
and
$$\triangle PQq = 8 \triangle PRQ.$$

Also, if RW, the ordinate from R to PV, be produced to meet the curve again in r,
$$RW = rW,$$
and the same proof shows that
$$\triangle PQq = 8 \triangle Prq.$$

Proposition 22. *If there be a series of areas $A, B, C, D, \ldots$ each of which is four times the next in order, and if the largest, A, be equal to the triangle PQq inscribed in a parabolic segment PQq and having the same base with it and equal height, then*
$$(A + B + C + D + \cdots) < (area\ of\ segment\ PQq).$$

For, since $\triangle PQq = 8 \triangle PRQ = 8 \triangle Pqr$, where R, r are the vertices of the segments cut off by PQ, Pq, as in the last proposition,
$$\triangle PQq = 4(\triangle PQR + \triangle Pqr).$$
Therefore, since $\triangle PQq = A$,
$$\triangle PQR + \triangle Pqr = B.$$

In like manner we prove that the triangles similarly inscribed in the remaining segments are together equal to the area C, and so on.

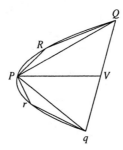

Therefore, $A + B + C + D + \cdots$ is equal to the area of a certain inscribed polygon, and is therefore less than the area of the segment.

Take areas $b, c, d, \ldots$ such that

$$b = \frac{1}{3}B,$$

$$c = \frac{1}{3}C,$$

$$d = \frac{1}{3}D, \quad \text{and so on.}$$

Then, since

$$b = \frac{1}{3}B,$$

and

$$B = \frac{1}{4}A,$$

$$B + b = \frac{1}{3}A.$$

Similarly

$$C + c = \frac{1}{3}B.$$

$$\cdots$$

Therefore

$$B + C + D + \cdots + Z + b + c + d + \cdots + z = \frac{1}{3}(A + B + C + \cdots + Y).$$

But

$$b + c + d + \ldots + y = \frac{1}{3}(B + C + D + \cdots + Y).$$

Therefore, by subtraction

$$B + C + D + \cdots + Z + z = \frac{1}{3}A$$

or

$$A + B + C + \cdots + Z + \frac{1}{3}Z = \frac{4}{3}A.$$

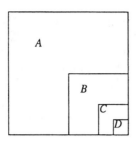

[The algebraic equivalent of this result is of course

$$1 + \frac{1}{4} + \left(\frac{1}{4}\right)^2 + \cdots + \left(\frac{1}{4}\right)^{n-1} = \frac{4}{3} - \frac{1}{3}\left(\frac{1}{4}\right)^{n-1}$$

$$= \frac{1 - \left(\frac{1}{4}\right)^n}{1 - \frac{1}{4}}.]$$

Proposition 24. *Every segment bounded by a parabola and a chord Qq is equal to four-thirds of the triangle which has the same base as the segment and equal height.*

Suppose

$$K = \frac{4}{3} \triangle PQq,$$

where P is the vertex of the segment; and we have then to prove that the area of the segment is equal to K.

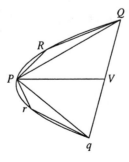

For, if the segment be not equal to K, it must either be greater or less.

I. Suppose the area of the segment is greater than K.

If then we inscribe in the segments cut off by PQ, Pq triangles which have the same base and equal height, i.e., triangles with the same vertices R, r as those of the segments, and if in the remaining segments we inscribe triangles in the same manner, and so on, we shall finally have segments remaining whose sum is less than the area by which the segment PQq exceeds K.

Therefore the polygon so formed must be greater than the area K; which is impossible, since [Prop. 23]

$$A + B + C + \cdots + Z < \frac{4}{3}A,$$

where

$$A = \triangle PQq.$$

Thus the area of the segment cannot be greater than K.

II. Suppose, if possible, that the area of the segment is less than K.

If then $\triangle PQq = A$, $B = \frac{1}{4}A$, $C = \frac{1}{4}B$, and so on, until we arrive at an area X such that X is less than the difference between K and the segment, we have

$$A + B + C + \cdots + X + \frac{1}{3}X = \frac{4}{3}A \qquad \text{[Prop. 23]}$$
$$= K$$

Now, since K exceeds $A + B + C + \cdots + X$ by an area less than X, and the area of the segment by an area greater than X, it follows that

$$A + B + C + \cdots + X > \text{(the segment)};$$

which is impossible, by Prop. 22 above.

Hence the segment is not less than K.

Thus, since the segment is neither greater nor less than K,

$$\text{(area of segment } PQq) = K = \frac{4}{3} \triangle PQq.$$

Appendix B

On a Method for the
Evaluation of Maxima and Minima
by Pierre de Fermat

The whole theory of evaluation of maxima and minima presupposes two unknown quantities and the following rule:

Let a be any unknown of the problem (which is in one, two, or three dimensions, depending on the formulation of the problem). Let us indicate the maximum or minimum by a in terms which could be of any degree. We shall now replace the original unknown a by $a + e$ and we shall express thus the maximum or minimum quantity in terms of a and e involving any degree. We shall adequate [adégaler], to use Diophantus' term, the two expressions of the maximum or minimum quantity and we shall take out their common terms. Now it turns out that both sides will contain terms in e or its powers. We shall divide all terms by e, or by a higher power of e, so that e will be completely removed from at least one of the terms. We suppress then all the terms in which e or one of its powers will still appear, and we shall equate

From a letter sent to Father Marin Mersenne who then forwarded it to Descartes. Reprinted, with permission, from *A Source Book in Mathematics 1200-1800*, D. J. Struik, ed. (Princeton, New Jersey: Princeton University Press, 1986).

the others; or, if one of the expressions vanishes, we shall equate, which is the same thing, the positive and negative terms. The solution of this last equation will yield the value of a, which will lead to the maximum or minimum, by using again the original expression.

Here is an example:

To divide the segment AC [Fig. 1] *at E so that AE × EC may be a maximum.*

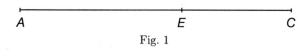

$$A \qquad\qquad\qquad E \qquad\qquad C$$

Fig. 1

We write $AC = b$; Let a be one of the segments, so that the other will be $b - a$, and the product, the maximum of which is to be found, will be $ba - a^2$. Let now $a + e$ be the first segment of b; the second will be $b - a - e$, and the product of the segments, $ba - a^2 + be - 2ae - e^2$; this must be adequated with the preceding: $ba - a^2$. Suppressing common terms: $be \sim 2ae + e$. Suppressing e: $b = 2a$. To solve the problem we must consequently take the half of b.

We can hardly expect a more general method.

Appendix C

From a Letter to Henry Oldenburg
on the Binomial Series (June 13, 1676)
by Isaac Newton

Most worthy Sir,

Though the modesty of Mr. Leibniz, in the extracts from his letter which you have lately sent me, pays great tribute to our countrymen for a certain theory of infinite series, about which there now begins to be some talk, yet I have no doubt that he has discovered not only a method for reducing any quantities whatever to such series, as he asserts, but also various shortened forms, perhaps like our own, if not even better. Since, however, he very much wants to know what has been discovered in this subject by the English, and since I myself fell upon this theory some years ago, I have sent you some of those things which occurred to me in order to satisfy his wishes, at any rate in part.

Fractions are reduced to infinite series by division; and radical quantities by extraction of the roots, by carrying out those operations in the symbols

Reprinted, with permission, from *The Correspondence of Isaac Newton*, H. W. Turnbull, F. R. S., ed. Cambridge University Press, Cambridge, United Kingdom, 1960.

just as they are commonly carried out in decimal numbers. These are the foundations of these reductions: but extractions of roots are much shortened by this theorem

$$(P + PQ)^{m/n} = P^{m/n} + \frac{m}{n}AQ + \frac{m-n}{2n}BQ + \frac{m-2n}{3n}CQ$$
$$+ \frac{m-3n}{4n}DQ + \text{etc.,}$$

where $P + PQ$ signifies the quantity whose root or even any power, or the root of a power, is to be found; P signifies the first term of that quantity, Q the remaining terms divided by the first, and m/n the numerical index of the power of $P + PQ$, whether that power is integral or (so to speak) fractional, whether positive or negative. For as analysts, instead of aa, aaa, etc., are accustomed to write a^2, a^3, etc., so instead of $\sqrt{a}$, $\sqrt{a^3}$, $\sqrt{c:a^5}$, etc. I write $a^{\frac{1}{2}}$ $a^{\frac{3}{2}}$, $a^{\frac{5}{3}}$, and instead of $1/a$, $1/aa$, $1/a^3$, I write a^{-1}, a^{-2}, a^{-3}. And so for

$$\frac{aa}{\sqrt{c:(a^3 + bbx)}}$$

I write $aa(a^3 + bbx)^{-\frac{1}{2}}$, and for

$$\frac{aab}{\sqrt{c:\{(a^3 + bbx)(a^3 + bbx)\}}}$$

I write $aab(a^3 + bbx)^{-\frac{2}{3}}$: in which last case, if $(a^3 + bbx)^{-\frac{2}{3}}$ is supposed to be $(P + PQ)^{m/n}$ in the Rule, then P will be equal to a^3, Q to bbx/a^3, m to -2, and n to 3. Finally, for the terms found in the quotient in the course of the working I employ A, B, C, D, etc., namely, A for the first term, $P^{m/n}$; B for the second term, $(m/n)AQ$; and so on. For the rest, the use of the rule will appear from the examples.

Example 1.

$$\sqrt{(c^2 + x^2)} \quad \text{or} \quad (c^2 + x^2)^{\frac{1}{2}} = c + \frac{x^2}{2c} - \frac{x^4}{8c^3} + \frac{x^6}{16c^5} - \frac{5x^8}{128c^7} + \frac{7x^{10}}{256c^9} + \text{etc.}$$

For in this case, $P = c^2$, $Q = x^2/c^2$, $m = 1$, $n = 2$, $A\left(= P^{m/n} = (cc)^{\frac{1}{2}}\right) = c$, $B\left(= (m/n)AQ\right) = x^2/2c$, $C\left(= \frac{m-n}{2n}BQ\right) = -\frac{x^4}{8c^3}$; and so on.

Appendix D

From a Letter to Henry Oldenburg on the Binomial Series (October 24, 1676) by Isaac Newton

Most worthy Sir,

I can hardly tell with what pleasure I have read the letters of those very distinguished men Leibniz and Tschirnhaus. Leibniz's method for obtaining convergent series is certainly very elegant, and it would have sufficiently revealed the genius of its author, even if he had written nothing else. But what he has scattered elsewhere throughout his letter is most worthy of his reputation—it leads us also to hope for very great things from him. The variety of ways by which the same goal is approached has given me the greater pleasure, because three methods of arriving at series of that kind had already become known to me, so that I could scarcely expect a new one to be communicated to us. One of mine I have described before; I now add another, namely, that by which I first chanced on these series—for I chanced on them before I knew the divisions and extractions of roots which I now use. And an

Reprinted, with permission, from *The Correspondence of Isaac Newton*, H. W. Turnbull, F. R. S., ed. Cambridge University Press, Cambridge, United Kingdom, 1960.

explanation of this will serve to lay bare, what Leibniz desires from me, the basis of the theorem set forth near the beginning of the former letter.

At the beginning of my mathematical studies, when I had met with the works of our celebrated Wallis, on considering the series by the intercalation of which he himself exhibits the area of the circle and the hyperbola, the fact that, in the series of curves whose common base or axis is x and the ordinates

$$(1-x^2)^{\frac{0}{2}}, \quad (1-x^2)^{\frac{1}{2}}, \quad (1-x^2)^{\frac{2}{2}}, \quad (1-x^2)^{\frac{3}{2}}, \quad (1-x^2)^{\frac{4}{2}}, \quad (1-x^2)^{\frac{5}{2}}, \quad \text{etc.},$$

if the areas of every other of them, namely

$$x, \quad x - \frac{1}{3}x^3, \quad x - \frac{2}{3}x^3 + \frac{1}{5}x^5, \quad x - \frac{3}{3}x^3 + \frac{3}{5}x^5 - \frac{1}{7}x^7, \quad \text{etc.,}$$

could be interpolated, we should have the areas of the intermediate ones, of which the first $(1 - x^2)^{\frac{1}{2}}$ is the circle: in order to interpolate these series I noted that in all of them the first term was x and that the second terms $\frac{0}{3}x^3, \frac{1}{3}x^3, \frac{2}{3}x^3, \frac{3}{3}x^3$, etc., were in arithmetical progression, and hence that the first two terms of the series to be intercalated ought to be $x - \frac{1}{3}(\frac{1}{2}x^3), x - \frac{1}{3}(\frac{3}{2}x^3), x - \frac{1}{3}(\frac{5}{2}x^3)$, etc. To intercalate the rest I began to reflect that the denominators 1, 3, 5, 7, etc. were in arithmetical progression, so that the numerical coefficients of the numerators only were still in need of investigation. But in the alternately given areas these were the figures of powers of the number 11, namely of these, $11^0, 11^1, 11^2, 11^3, 11^4$, that is, first 1; then 1, 1; thirdly, 1, 2, 1; fourthly 1, 3, 3, 1; fifthly 1, 4, 6, 4, 1, etc. An so I began to inquire how the remaining figures in these series could be derived from the first two given figures, and I found that on putting m for the second figure, the rest would be produced by continual multiplication of the terms of this series,

$$\frac{m-0}{1} \times \frac{m-1}{2} \times \frac{m-2}{3} \times \frac{m-3}{4} \times \frac{m-4}{5}, \quad \text{etc.}$$

For example, let $m = 4$, and $4 \times \frac{1}{2}(m - 1)$, that is 6 will be the third term, and $6 \times \frac{1}{3}(m - 2)$, that is 4 the fourth, and $4 \times \frac{1}{4}(m - 3)$, that is 1 the fifth, and $1 \times \frac{1}{5}(m - 4)$, that is 0 the sixth, at which term in this case the series stops. Accordingly, I applied this rule for interposing series among series, and since, for the circle, the second term was $\frac{1}{3}(\frac{1}{2}x^3)$, I put $m = \frac{1}{2}$, and the terms arising were

$$\frac{1}{2} \times \frac{\frac{1}{2}-1}{2} \quad \text{or} \quad -\frac{1}{8}, \qquad -\frac{1}{8} \times \frac{\frac{1}{2}-2}{3} \quad \text{or} \quad +\frac{1}{16}, \qquad \frac{1}{16} \times \frac{\frac{1}{2}-3}{4} \quad \text{or} \quad -\frac{5}{128},$$

and so to infinity. Whence I came to understand that the area of the circular segment which I wanted was

$$x - \frac{\frac{1}{2}x^3}{3} - \frac{\frac{1}{8}x^5}{5} - \frac{\frac{1}{16}x^7}{7} - \frac{\frac{1}{128}x^9}{9} \quad \text{etc.}$$

And by the same reasoning the areas of the remaining curves, which were to be inserted, were likewise obtained: as also the area of the hyperbola and of the other alternate curves in this series $(1 + x^2)^{\frac{0}{2}}$, $(1 + x^2)^{\frac{1}{2}}$, $(1 + x^2)^{\frac{2}{2}}$, $(1 + x^2)^{\frac{3}{2}}$, etc. And the same theory serves to intercalate other series, and that through intervals of two or more terms when they are absent at the same time. This was my first entry upon these studies, and it had certainly escaped my memory, had I not a few weeks ago cast my eye back on some notes.

But when I had learnt this, I immediately began to consider that the terms

$$(1 - x^2)^{\frac{0}{2}}, \quad (1 - x^2)^{\frac{2}{2}}, \quad (1 - x^2)^{\frac{4}{2}}, \quad (1 - x^2)^{\frac{6}{2}}, \quad \text{etc.,}$$

that is to say,

$$1, \quad 1 - x^2, \quad 1 - 2x^2 + x^4, \quad 1 - 3x^2 + 3x^4 - x^6, \quad \text{etc.}$$

could be interpolated in the same way as the areas generated by them: and that nothing else was required for this purpose but to omit the denominators 1, 3, 5, 7, etc., which are in the terms expressing the areas; this means that the coefficients of the terms of the quantity to be intercalated $(1 - x^2)^{\frac{1}{2}}$, or $(1 - x^2)^{\frac{3}{2}}$, or in general $(1 - x^2)^m$, arise by the continued multiplication of the terms of this series

$$m \times \frac{m - 1}{2} \times \frac{m - 2}{3} \times \frac{m - 3}{4}, \quad \text{etc.,}$$

so that (for example)

$$(1 - x^2)^{\frac{1}{2}} \quad \text{was the value of } 1 - \frac{1}{2}x^2 - \frac{1}{8}x^4 - \frac{1}{16}x^6, \quad \text{etc.,}$$

$$(1 - x^2)^{\frac{3}{2}} \quad \text{of } 1 - \frac{3}{2}x^2 + \frac{3}{8}x^4 + \frac{1}{16}x^6, \quad \text{etc.,}$$

and

$$(1 - x^2)^{\frac{1}{3}} \quad \text{of } 1 - \frac{1}{3}x^2 - \frac{1}{9}x^4 - \frac{5}{81}x^6, \quad \text{etc.}$$

So then the general reduction of radicals into infinite series by that rule, which I laid down at the beginning of my earlier letter, became known to me, and that before I was acquainted with the extraction of roots. But once this was known, that other could not long remain hidden from me. For in order to test these processes, I multiplied

$$1 - \frac{1}{2}x^2 - \frac{1}{8}x^4 - \frac{1}{16}x^6, \quad \text{etc.}$$

into itself; and it became $1 - x^2$, the remaining terms vanishing by the continuation of the series to infinity. And even so $1 - \frac{1}{3}x^2 - \frac{1}{9}x^4 - \frac{5}{81}x^6$, etc. multiplied twice into itself also produced $1 - x^2$. And as this was not only sure proof of these conclusions so too it guided me to try whether, conversely, these series, which it thus affirmed to be roots of the quantity $1 - x^2$, might not be extracted out of it in an arithmetical manner. And the matter turned out well. This was the form of the working in square roots.

$$1 - x^2(1 - \frac{1}{2}x^2 - \frac{1}{8}x^4 - \frac{1}{16}x^6, \quad \text{etc.}$$

$$1$$

$$0 - x^2$$

$$- x^2 + \frac{1}{4}x^4$$

$$- \frac{1}{4}x^4$$

$$- \frac{1}{4}x^4 + \frac{1}{8}x^6 + \frac{1}{64}x^8$$

$$0 \quad - \frac{1}{8}x^6 - \frac{1}{64}x^8$$

Appendix E

Excerpts from "Of Analysis by Equations of an Infinite Number of Terms" by Isaac Newton

1. The General Method, which I had devised some considerable Time ago, for measuring the Quantity of Curves, by Means of Series, infinite in the Number of Terms, is rather shortly explained, than accurately demonstrated in what follows.

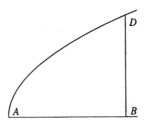

Reprinted, with permission, from *The Mathematical Works of Isaac Newton*, D. T. Whiteside, ed. (New York: Johnson Reprint Corporation, 1964). (Translated in 1745 by John Stewart.)

2. Let the Base AB of any Curve AD have BD for it's perpendicular Ordinate; and call $AB = x$, $BD = y$, and let a, b, c, *Etc.* be given Quantities, and m and n whole Numbers. Then

The Quadrature of Simple Curves.
RULE I.

3. If $ax^{m/n} = y$; it shall be $\dfrac{an}{m+n}x^{(m+n)/n} = $ Area ABD.

The thing will be evident by an Example.

1. If $x^2(= 1x^{2/1}) = y$, that is $a = 1 = n$, and $m = 2$; it shall be $\frac{1}{3}x^3 = ABD$.

2. Suppose $4\sqrt{x}(= 4x^{1/2}) = y$; it will be $\frac{8}{3}x^{3/2}\left(= \frac{8}{3}\sqrt{x^3}\right) = ABD$.

3. If $\sqrt[3]{x^5}(= x^{5/3}) = y$; it will be $\frac{3}{8}x^{8/3}\left(= \frac{3}{8}\sqrt[3]{x^8}\right) = ABD$.

4. If $1/x^2(= x^{-2}) = y$, that is if $a = 1 = n$, and $m = -2$;

It will be $\frac{1}{-1}x^{-1/1} = -x^{-1}(= -1/x) = \alpha BD$, infinitely extended towards α, which the Calculation places negative, because it lyes upon the other side of the Line BD.

5. If $\frac{1}{\sqrt{x^3}}(x^{-3/2}) = y$; it will be $\left(\frac{2}{-1}x^{-1/2} = \right)\frac{2}{-\sqrt{x}} = BD\alpha$.

6. If $\frac{1}{x}(= x^{-1}) = y$; it will be $\frac{1}{0}x^{0/1} = \frac{1}{0}x^0 = \frac{1}{0} \times 1 = \frac{1}{0} = $ an infinite Quantity; such as is the Area of the Hyperbola upon both Sides of the Line BD.

The Quadrature of Curves compounded of simple ones.
RULE II.

4. If the Value of y be made up of several such Terms, the Area likewise shall be made up of the Areas which result from every one of the Terms.

The first Examples.

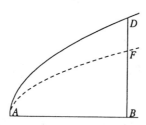

5. If it be $x^2 + x^{3/2} = y$; it will be $\frac{1}{3}x^3 + \frac{2}{5}x^{5/2} = ABD$.

For if it be always $x^2 = BF$ and $x^{3/2} = FD$, you will have by the preceding Rule $\frac{1}{3}x^3 =$ Superficies AFB; described by the Line BF; and $\frac{2}{5}x^{5/2} = AFD$ described by DF; wherefore $\frac{1}{3}x^3 + \frac{2}{5}x^{5/2} =$ the whole Area ABD.

Thus if it be $x^2 - x^{3/2} = y$; it will be $\frac{1}{3}x^3 - \frac{2}{5}x^{5/2} = ABD$. And if it be $3x - 2x^2 + x^3 - 5x^4 = y$; it will be $\frac{3}{2}x^2 - \frac{2}{3}x^3 + \frac{1}{4}x^4 - x^5 = ABD$.

The second Examples.

6. If $x^{-2} + x^{-3/2} = y$; it will be $-x^{-1} - 2x^{-1/2} = \alpha BD$. Or if it be $x^{-2} - x^{-3/2} = y$; it will be $-x^{-1} + 2x^{-1/2} = \alpha BD$.

And if you change the Signs of the Quantities, you will have the affirmative Value $(x^{-1} + 2x^{-1/2}$, or $x^{-1} - 2x^{-1/2})$ of the Superficies αBD, provided the whole of it fall above the Base $AB\alpha$.

7. But if any Part fall below (which happens when the Curve decussates or crosses it's Base betwixt B and α, as you see here in δ) you are to subtract that Part from the Part above the Base; and so you shall have the Value of

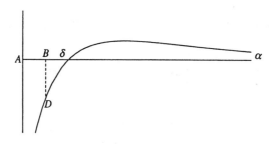

the Difference: but if you would have their Sum, seek both the Superficies's separately, and add them. And the same thing I would have observed in the other Examples belonging to this Rule.

The third Examples.

8. If $x^2 + x^{-2} = y$; it will be $\frac{1}{3}x^3 - x^{-1} =$ the Superficies described.

But here it must be remarked that the Parts of the said Superficies so found, lye upon opposite Sides of the Line BD.

That is, putting $x^2 = BF$, and $x^{-2} = FD$; it shall be $\frac{1}{3}x^3 = ABF$ the Superficies described by BF, and $-x^{-1} = DF\alpha$ the Superficies described by DF.

9. And this always happens when the Indexes $\left(\frac{m \pm n}{n}\right)$ of the Ratios of the Base x in the Value of the Superficies sought, are affected with different Signs.

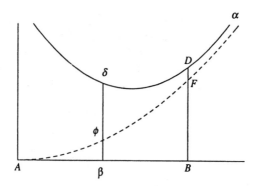

In such Cases any middle part $BD\delta\beta$ of the Superficies (which only can be given, when the Superficies is infinite upon both Sides) is thus found.

Substract the Superficies belonging to the lesser Base $A\beta$ from the Superficies belonging to the greater Base AB, and you shall have $\beta BD\delta$ the Superficies insisting upon the difference of the Bases. Thus in this Example (see the preceding Fig.)

If $AB = 2$, and $A\beta = 1$; it will be $\beta BD\delta = 17/6$:

For the Superficies belonging to AB (viz. $ABF - DF\alpha$) will be $8/3 - 1/2$ or $13/6$, and the Superficies belonging to $A\beta$ (viz. $A\phi\beta - \delta\phi\alpha$) will be $1/3 - 1$, or $-2/3$: and their Difference (viz. $ABF - DF\alpha - A\phi\beta + \delta\phi\alpha - \beta BD\delta$) will be $13/6 + 2/3$ or $17/6$.

After the same manner, if $A\beta = 1$, and $AB = x$; it will be $\beta BD\delta = \frac{2}{3} + \frac{1}{3}x^3 - x^{-1}$.

Thus if $2x^3 - 3x^5 - \frac{2}{3}x^{-4} + x^{-3/5} = y$, and $A\beta = 1$;

It will be $\beta BD\delta = \frac{1}{2}x^4 - \frac{1}{2}x^6 + \frac{2}{9}x^{-3} + \frac{5}{2}x^{2/5} - \frac{49}{18}$.

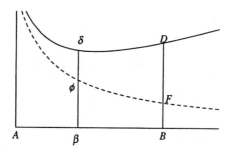

10. Finally it may be observed, that if the Quantity x^{-1} be found in the Value of y, that Term (since it generates an hyperbolical Surface) is to be considered apart from the rest.

As if it were $x^2 + x^{-3} + x^{-1} = y$: let it be $x^{-1} = BF$, and $x^2 + x^{-3} = FD$ and $A\beta = 1$; and it will be $\delta\phi FD = \frac{1}{6} + \frac{1}{3}x^3 - \frac{1}{2}x^{-2}$, as being that which is generated by the Terms $x^2 + x^{-3}$.

Wherefore if the remaining Superficies $\beta\phi FB$, which is hyperbolical, be given by any Method of Computation, the whole βBDd will be given.

The Quadrature of all other Curves.
RULE III.

11. But if the Value of y, or any of it's Terms be more compounded than the foregoing, it must be reduced into more simple Terms; by performing the Operation in Letters, after the same Manner as Arithmeticians divide in Decimal Numbers, extract the Square Root, or resolve affected Equations; and afterwards by the preceding Rules you will discover the Superficies of the Curve sought.

Examples, where you divide.

12. Let $aa/(b + x) = y$; *Viz.* where the Curve is an Hyperbola. Now that that Equation may be freed from it's Denominator, I make the Division thus.

$$b + x)aa + 0 \left(\frac{aa}{b} - \frac{aax}{b^2} + \frac{aax^2}{b^3} - \frac{aax^3}{b^4} \right. \; Etc.$$

$$aa + \frac{aax}{b}$$

$$\overline{}$$

$$0 - \frac{aax}{b} + 0$$

$$- \frac{aax}{b} - \frac{aax^2}{b^2}$$

$$\overline{}$$

$$0 + \frac{aax^2}{b^2} + 0$$

$$+ \frac{aax^2}{b^2} + \frac{aax^3}{b^3}$$

$$\overline{}$$

$$0 - \frac{aax^3}{b^3} + 0$$

$$- \frac{aax^3}{b^3} - \frac{aax^3}{b^4}$$

$$\overline{}$$

$$0 + \frac{aax^4}{b^4} \quad Etc.$$

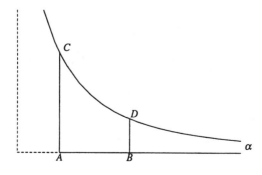

And thus in Place of this $y = aa/(b + x)$, a new Equation arises, *viz.*
$y = \frac{a^2}{b} - \frac{a^2 x}{b^2} + \frac{a^2 x^2}{b^3} - \frac{a^2 x^3}{b^4}$ *Etc.* this Series being continued infinitely; and therefore (by the second Rule)

The Area sought $ABDC$ will be equal to $\frac{a^2 x}{b} - \frac{a^2 x^2}{2b^2} + \frac{a^2 x^3}{3b^3} - \frac{a^2 x^4}{4b^4}$ *Etc.* an infinite Series likewise, but yet such, that a few of the initial Terms are exact enough for any Use, provided that b be equal to x repeated some few times.

13. After the same Manner if it be $1/(1 + xx) = y$, by dividing there arises $y = 1 - xx + x^4 - x^6 + x^8$ *Etc.* Whence (by the second Rule)

You will have $ABDC = x - \frac{1}{3}x^3 + \frac{1}{5}x^5 - \frac{1}{7}x^7 + \frac{1}{9}x^9$ *Etc.*

Or if x^2 be made the first Term in the Divisor, *viz.* thus: $x^2 + 1$) there will arise $x^{-2} - x^{-4} + x^{-6} - x^{-8}$ *Etc.* for the Value of y; whence (by the second Rule)

It will be $BD\alpha = -x^{-1} + \frac{1}{3}x^{-3} - \frac{1}{5}x^{-5} + \frac{1}{7}x^{-7}$ *Etc.* You must proceed in the first Way when x is small enough, but the second Way, when it is supposed great enough.

14. Finally, if it be $\dfrac{2x^{1/2} - x^{3/2}}{1 + x^{1/2} - 3x} = y$; by dividing there arises $2x^{1/2} - 2x + 7x^{3/2} - 13x^2 + 34x^{5/2}$ *Etc.* whence it will be $ABDC = \frac{4}{3}x^{3/2} - x^2 + \frac{14}{5}x^{5/2} - \frac{13}{3}x^3$ *Etc.*

Examples, where the Square Root must be extracted.

15. If it be $\sqrt{aa + xx} = y$, I extract the Root thus:

$$aa + xx \left(a + \frac{x^2}{2a} - \frac{x^4}{8a^3} + \frac{x^6}{16a^5} - \frac{5x^8}{128a^7} \; Etc. \right.$$

aa

$\overline{}$

$0 + x^2$

$\quad x^2 + \dfrac{x^4}{4a^2}$

$\overline{}$

$\quad 0 - \dfrac{x^4}{4a^2}$

$\quad\quad - \dfrac{x^4}{4a^2} - \dfrac{x^6}{8a^4} + \dfrac{x^8}{64a^6}$

$\overline{}$

$\quad\quad 0 + \dfrac{x^6}{8a^4} - \dfrac{x^8}{64a^6}$

$\quad\quad\quad + \dfrac{x^6}{8a^4} + \dfrac{x^8}{16a^6} - \dfrac{x^{10}}{64a^8} + \dfrac{x^{12}}{256a^{10}}$

$\overline{}$

$\quad\quad\quad 0 - \dfrac{5x^8}{64a^6} + \dfrac{x^{10}}{64a^8} - \dfrac{x^{12}}{256a^{10}} \quad Etc.$

Whence for the Equation $\sqrt{aa + xx} = y$, a new one is produced, *vis.* $y = a + \frac{x^2}{2a} - \frac{x^4}{8a^3} + \frac{x^6}{16a^5} - \frac{5x^8}{128a^7}$ *Etc.* And (by the second Rule)

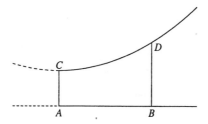

You will have the Area sought $ABDC = ax + \frac{x^3}{6a} - \frac{x^5}{40a^3} + \frac{x^7}{112a^5} - \frac{5x^9}{1152a^7}$ Etc.

And this is the quadrature of the Hyperbola.

16. After the same Manner if it be $\sqrt{aa - xx} = y$, it's Root will be $a - \frac{x^2}{2a} - \frac{x^4}{8a^3} - \frac{x^6}{16a^5} - \frac{5x^8}{128a^7}$ Etc. and therefore the Area sought $ABDC$ will be equal to $ax - \frac{x^3}{6a} - \frac{x^5}{40a^3} - \frac{x^7}{112a^5} - \frac{5x^9}{1152a^7}$ Etc. And this is the quadrature of the Circle.

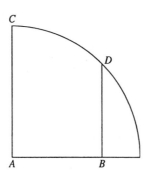

17. Or if you suppose $\sqrt{x - xx} = y$, the Root will be equal to this infinite Series

$$x^{1/2} - \frac{1}{2}x^{3/2} - \frac{1}{8}x^{5/2} - \frac{1}{16}x^{7/2} - \frac{5}{128}x^{9/2} \text{ Etc.}$$

And the Area sought ABD will be

$$\frac{2}{3}x^{3/2} - \frac{1}{5}x^{5/2} - \frac{1}{28}x^{7/2} - \frac{1}{72}x^{9/2} - \frac{5}{704}x^{11/2} \text{ Etc.}$$

Or

$$x^{1/2}\left(\frac{2}{3}x - \frac{1}{5}x^2 - \frac{1}{28}x^3 - \frac{1}{72}x^4 - \frac{5}{704}x^5 \text{ Etc.}\right)$$

18. If $\dfrac{\sqrt{1 + ax^2}}{\sqrt{1 - bx^2}} = y$ (whose Quadrature gives the Length of the Curve of the Ellipse) by extracting both Roots, there arises

$$\frac{1 + \frac{1}{2}ax^2 - \frac{1}{8}a^2x^4 + \frac{1}{16}a^3x^6 - \frac{5}{128}a^4x^8}{1 - \frac{1}{2}bx^2 - \frac{1}{8}b^2x^4 - \frac{1}{16}b^3x^6 - \frac{5}{128}b^4x^8} \text{ Etc.}$$

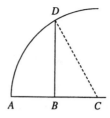

And by dividing as is done in the Case of Decimal Fractions, you will have

$$1+\frac{1}{2}bx^2+\frac{1}{8}b^2x^4+\frac{5}{16}b^3x^6+\frac{35}{128}b^4x^8 \ Etc.$$

$$+\frac{1}{2}a \quad +\frac{1}{4}ab \quad +\frac{3}{16}ab^2 \quad +\frac{5}{32}ab^3$$

$$-\frac{1}{8}a^2 \quad -\frac{1}{16}a^2b \quad -\frac{3}{64}a^2b^2$$

$$+\frac{1}{16}a^3 \quad +\frac{1}{32}a^3b$$

$$-\frac{5}{128}a^4$$

And therefore the Area required

$$x+\frac{1}{6}bx^3+\frac{3}{40}b^2x^5 \ Etc.$$

$$+\frac{1}{6}a \quad +\frac{1}{20}ab$$

$$-\frac{1}{40}a^2$$

19. But it is to be observed that the Operation may be often abbreviated by a due Preparation of the Equation, as in the Example just now adduced $\frac{\sqrt{1+ax^2}}{\sqrt{1-bx^2}}=y$. If you multiply both Parts of the Fraction by $\sqrt{1-bx^2}$, there will arise $\frac{\sqrt{1+(a-b)x^2-abx^4}}{1-bx^2}=y$; and the rest of the work is performed by extracting the Root of the Numerator only, and then dividing by the Denominator.

20. From these Examples, I suppose, it will be sufficiently evident after what Manner any Value of y may be reduced (with whatever Roots or Denominators it may be involved, as you may see here

$$x^3 + \frac{\sqrt{x - \sqrt{1 - xx}}}{\sqrt[3]{axx + x^3}} - \frac{\sqrt[5]{x^3 + 2x^5 - x^{3/2}}}{\sqrt[5]{x + x^2} - \sqrt{2x - x^{2/3}}} = y\Bigg)$$

into infinite Series of simple Terms, from which, by the second Rule, the Superficies required may be known.

* * *

52. Neither do I know any Thing of this Kind to which this Method doth not extend, and that in various Ways. Yea Tangents may be drawn to Mechanical Curves by it, when it happens that it can be done by no other Means. And whatever the common Analysis performs by Means of Equations of a finite Number of Terms (provided that can be done) this can always perform the same by Means of infinite Equations: So that I have not made any Question of giving this the Name of *Analysis* likewise. For the Reasonings in this are no less certain than in the other; nor the Equations less exact; albeit we Mortals whose reasoning Powers are confined within narrow Limits, can neither express, nor so conceive all the Terms of these Equations, as to know exactly from thence the Quantities we want: Even as the surd Roots of finite Equations can neither be so exprest by Numbers, nor any analytical Contrivance, that the Quantity of any one of them can be so distinguished from all the rest, as to be understood exactly.

53. To conclude, we may justly reckon that to belong to the *Analytic* Art, by the Help of which the Areas and Lengths *Etc.* of Curves may be exactly and geometrically determined (when such a thing is possible). But this is not a Place for insisting upon these Things. There are two Things especially which an attentive Reader will see need to be demonstrated.

I. The Demonstration of the Quadrature of Simple Curves belonging to Rule the first.

Preparation for demonstrating the first Rule.

54. Let then $AD\delta$ be any Curve whose Base $AB = x$, the perpendicular Ordinate $BD = y$, and the Area $ABD = z$, as at the Beginning. Likewise put $B\beta = o$, $BK = v$; and the Rectangle $B\beta HK(ov)$ equal to the Space $B\beta\delta D$.

Therefore it is $A\beta = x + o$, and $A\delta\beta = z + ov$: Which Things being premised, assume any Relation betwixt x and z that you please, and seek for y in the following Manner.

Take at Pleasure $\frac{2}{3}x^{3/2} = z$; or $\frac{4}{9}x^3 = z^2$. Then $x + o(A\beta)$ being substituted for x, and $z + ov(A\delta\beta)$ for z, there arises 4/9 into $x^3 + 3xo^2 + 3xo^2 + o^3 = $ (from the Nature of the Curve) $z^2 + 2zov + o^2v^2$. And taking away Equals $\left(\frac{4}{9}x^3\right.$ and z^2) and dividing the Remainders by o, there arises 4/9 into $3x^2 + 3xo + oo = 2zv + ovv$. Now if we suppose $B\beta$ to be diminished infinitely and to vanish, or o to be nothing, v and y, in that Case will be equal, and the Terms which are multiplied by o will vanish: So that there will remain $\frac{4}{9} \times 3x^2 = 2vz$, or $\frac{2}{3}x^2(= zy) = \frac{2}{3}x^{3/2}y$; or $x^{1/2}(= x^2/x^{3/2}) = y$. Wherefore conversely if it be $x^{1/2} = y$, it shall be $\frac{2}{3}x^{3/2} = z$.

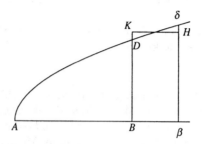

The Demonstration.

55. Or universally, if $\dfrac{n}{m+n} \times ax^{(m+n)/n} = z$; or, putting $\dfrac{na}{m+n} = c$, and $m + n = p$, if $cx^{p/n} = z$; or $c^n x^p = z^n$; Then by substituting $x + o$ for x, and $z + ov$ (or which is the same $z + oy$) for z, there arises c^n into $x^p + pox^{p-1}$, Etc.$= z^n + noyz^{n-1}$ Etc. the other Terms, which would at length vanish being neglected. Now taking away $c^n x^p$ and z^n which are equal, and dividing the Remainders by o, there remains

$$c^n p x^{p-1} = nyz^{n-1} \left(= \frac{nyz^n}{z} \right) = \frac{nyc^n x^p}{cx^{p/n}},$$

or, by dividing by $c^n x^p$, it shall be $px^{-1} = \dfrac{ny}{cx^{p/n}}$; or $pcx^{(p-n)/n} = ny$; or by restoring $\dfrac{na}{m+n}$ for c, and $m + n$ for p, that is m for $p - n$, and na for pc, it becomes $ax^{m/n} = y$. Wherefore conversely, if $ax^{m/n} = y$, it shall be $\dfrac{n}{m+n} ax^{(m+n)/n} = z$. Q.E.D.

To find those Curves which can be squared.

56. Hence by the Way you may observe after what Manner as many Curves as you please may be found, whose Areas are known; *viz.* assume any Equation you please for the Relation betwixt the Area z and Base x, and thence let the Ordinate y be sought. Thus if you suppose $\sqrt{aa + xx} = z$, by performing the Calculation you will find $\dfrac{x}{\sqrt{aa + xx}} = y$. And so in other Cases.

II. The Demonstration of the Resolution of Affected Equations.

57. The other Thing to be demonstrated, is the literal Resolution of affected Equations. *Viz.* That the Quotient, when x is sufficiently small, the further it is produced, approached so much nearer to the Truth, so that the defect (p, q, or r *Etc.*) by which it differs from the full Value of y, at Length becomes

less than any given Quantity; and that Quotient being produced infinitely is exactly equal to y. Which will appear thus.

1°. Because that Quantity in which x is the lowest Dimension (that is to say, more than the half of the last Term, provided you suppose x small enough) in every Operation is perpetually taken out of the last Term of the Equations, of which p, q, r, *Etc.* are the Roots: Therefore that last Term (by I. 10. Elem.) at length becomes less than any given Quantity; and would entirely vanish, if the Operation were infinitely continued.

Thus if it be $x = 1/2$, you have x the half of all these $x + x^2 + x^3 + x^4$, *Etc.* and x^2 the half of all these $x^2 + x^3 + x^4 + x^5$, *Etc.* Therefore if $x < 1/2$, x shall be greater than the half of all these $x + x^2 + x^3$, *Etc.* and x^2 greater than the half of all these $x^2 + x^3 + x^4$, *Etc.*

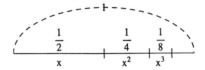

Thus if $x/b < 1/2$, you shall have x more than the half of all these $x + x^2/b + x^3/b^2$ *Etc.* And the same Way of others. And as to the numeral Coefficients, for most part they perpetually decrease: Or if at any Time they increase, you need only suppose x some few Times less.

2°. If the last Term of any Equation be continually diminished, till it at length vanish, one of it's Roots shall likewise be diminished, until the last Term vanishing, it vanish along with it.

3°. Wherefore one Value of the Quantities p, q, r, *Etc.* continually decreases until at length, when the Operation is infinitely produced, it vanish entirely.

4°. But the Values of these Roots p, q, r, *Etc.* together with the Quotient already extracted, are equal to the Roots of the proposed Equation (thus in the Resolution of the Equation $y^3 + a^2 y + axy - 2a^3 - x^3 = 0$, shewn above, you will observe that $y = a + p = a - \frac{1}{4}x + q = a - \frac{1}{4}x + \frac{xx}{64a} + r$, *Etc.*) Whence it is sufficiently evident that the Quotient infinitely produced, is one of the Values of y: Which was what I proposed to shew.

58. The same Thing will appear by substituting the Quotient instead of y in the Equation proposed. For you will perceive that those Terms perpetually destroy one another in which x is of the least Dimensions.

Appendix F

Excerpts from
"Subsiduum Calculi Sinuum"
by Leonhard Euler

Therefore, with the help of this theorem, the application of which is very broad, countless series can be produced following the sines and the cosines as they proceed from the multiples of any angle, and the sum of these series is established. I have only explained the case where the coefficients A, B, C, D, etc., proceed in geometric progression, but in the same way the calculation is adapted for other series. Furthermore, let it suffice to have noted that countless other series can be elicited from the series just found, by differentiation as well as by integration. For example, when there is

$$\cos \phi - \cos 2\phi + \cos 3\phi - \cos 4\phi + \cos 5\phi - \text{etc.} = \frac{1}{2}$$

by differentiating there will be

$$\sin \phi - 2 \sin 2\phi + 3 \sin 3\phi - 4 \sin 4\phi + 5 \sin 5\phi - \text{etc.} = 0$$

Opera (1) 14, 542–584. Translated by Wade Cartwright and printed with his permission.

and by differentiating again,

$$\cos\phi - 4\cos 2\phi + 9\cos 3\phi - 16\cos 4\phi + 25\cos 5\phi - \text{etc.} = 0$$

and so further.

However, the original series, multiplied by $d\phi$ and integrated, gives

$$\sin\phi - \frac{1}{2}\sin 2\phi + \frac{1}{3}\sin 3\phi - \frac{1}{4}\sin 4\phi + \frac{1}{5}\sin 5\phi - \text{etc} = \frac{\phi}{2}$$

where there is no need for the addition of a constant since, by setting $\phi = 0$, the sum automatically disappears. If this is multiplied by $-d\phi$ and again integrated, it will yield

$$\cos\phi - \frac{1}{4}\cos 2\phi + \frac{1}{9}\cos 3\phi - \frac{1}{16}\cos 4\phi + \text{etc.} = \alpha - \frac{\phi^2}{4}$$

and therefore, by setting $\phi = 0$,

$$1 - \frac{1}{4} + \frac{1}{9} - \frac{1}{16} + \text{etc.} = \alpha = \frac{\pi^2}{12}$$

as is established elsewhere.

Wherefore, if

$$\phi = \frac{\pi}{\sqrt{3}} = 103^\circ\ 55^{\mathrm{I}}\ 22^{\mathrm{II}}\ 58^{\mathrm{III}}\ 28^{\mathrm{IV}}$$

the sum of this series vanishes. However, lest I be excessively long, here I omit very many other outstanding variations of this type of series.

Solutions to Selected Exercises

Exercises 1.1

3. Achilles will overtake the tortoise after running $1,111\frac{1}{9}$ ft or $37\frac{1}{27}$ seconds.

5. 1

7. 1

8. a) 5/9   c) 23/180   e) 97/275

    g) 731/999000   i) 1231111/9999000

11. 30

Exercises 1.2

1. Let (x_c, y_c) denote the center of the chord that has equation $y = mx + c$. Then $x_c = -mcb^{-2}(a^{-2} + m^2 b^{-2})^{-1}$ and $y_c = mx_c + c$ and it is easily verified that y_c/x_c does not depend on c. It follows that as long as m is fixed (x_c, y_c) lies on a straight line through the origin.

3. Use the fact that at a maximum point the tangent to the curve is horizontal.

Exercises 2.1

1. a) $-1/2$ c) $1, -1$ e) $0, -2$

Exercises 3.1

1. a) $1 + \dfrac{x}{3} - \dfrac{x^2}{9} + \dfrac{5x^3}{81} - \dfrac{10x^4}{243} \ldots$ c) $1 + \dfrac{x^2}{3} - \dfrac{x^4}{9} \ldots$

e) $1 + \dfrac{2x}{3} - \dfrac{x^2}{9} + \dfrac{4x^3}{81} - \dfrac{7x^4}{243} \ldots$ g) $1 + \dfrac{2x^2}{3} - \dfrac{x^4}{9} \ldots$

i) $1 + \dfrac{5x}{3} + \dfrac{5x^2}{9} - \dfrac{5x^3}{81} + \dfrac{5x^4}{243} \ldots$ k) $1 + \dfrac{5x^2}{3} + \dfrac{5x^4}{9} \ldots$

m) $1 - 2x + 3x^2 - 4x^3 + 5x^4 \ldots$ o) $1 - 2x^2 + 3x^4 \ldots$

q) $1 - \dfrac{4x}{3} + \dfrac{14x^2}{9} - \dfrac{140x^3}{81} + \dfrac{455x^4}{243} \ldots$ s) $1 - \dfrac{4x^2}{3} + \dfrac{14x^4}{9} \ldots$

u) $1 + 3x + 6x^2 + 10x^3 + 15x^4 \ldots$

Exercises 3.2

1. a) $1 + 2x + 4x^2 + 8x^3 + 16x^4 \ldots$ c) $\dfrac{2}{3} - \dfrac{8x}{9} + \dfrac{32x^2}{27} - \dfrac{128x^3}{81} + \dfrac{512x^4}{243} \ldots$

e) $1 + x^2 + x^4 + x^6 + x^8 \ldots$ g) $1 + x + 9x^2 + 9x^3 + 81x^4 \ldots$

i) $\dfrac{1}{4} + \dfrac{x}{4} + \dfrac{x^2}{16} + \dfrac{x^3}{16} + \dfrac{x^4}{64} \ldots$ k) $1 + x - x^3 - x^4 + x^7 \ldots$

5. $\ln(1 + x) = x + x^2/2 + x^3/3 + x^4/4 + \ldots$

Exercises 3.3

1. a) Right-side $= \left(\dfrac{1}{2}\right)^{n+1} \left[1 + \dfrac{1}{2} + \left(\dfrac{1}{2}\right)^2 + \left(\dfrac{1}{2}\right)^3 \ldots\right]$

$= \left(\dfrac{1}{2}\right)^{n+1} \cdot \dfrac{1}{1 - \frac{1}{2}} = \left(\dfrac{1}{2}\right)^n.$

c) Right-side $\leq \dfrac{1}{2}\left(a_n x^n + a_n x^{n+1} + a_n x^{n+2} \ldots\right) = \dfrac{a_n}{2} x^n (1 + x + x^2 + \ldots)$

$< \dfrac{a_n}{2} x^n \left(1 + \dfrac{1}{2} + \left(\dfrac{1}{2}\right)^2 + \ldots\right) = \dfrac{a_n}{2} x^n \cdot 2 = a_n x^n$

Exercises 4.1

1. a) $y = c - cx + \dfrac{(1+c)x^2}{2} - \dfrac{(1+c)x^3}{6} + \dfrac{(1+c)x^4}{24} - \dfrac{(1+c)x^5}{120} \cdots$

c) $y = c + 2x + \dfrac{(-1+c)x^2}{2} + \dfrac{2x^3}{3} + \dfrac{(-1+c)x^4}{8} + \dfrac{2x^5}{15} \cdots$

e) $y = c + x + \dfrac{cx^2}{2} + \dfrac{x^3}{3} + \dfrac{cx^4}{8} + \dfrac{x^5}{15} \cdots$

g) $y = c + (1+c)x + \dfrac{(2+c)x^2}{2} + \dfrac{(4+c)x^3}{6} + \dfrac{(10+c)x^4}{24} + \dfrac{(34+c)x^5}{120} \cdots$

i) $y = c + (1+c)x + \dfrac{(3+2c)x^2}{4} + \dfrac{(5+4c)x^3}{24} + \dfrac{(13+8c)x^4}{192} + \dfrac{(11+16c)x^5}{1920} \cdots$

k) $y = -1 + cx$ **m)** $y = 2 + cx + \dfrac{3x^2}{2} + \dfrac{(-2+2c)x^3}{3} + \dfrac{x^4}{2} + \dfrac{(-3+2c)x^5}{5}$

o) $y = x - \dfrac{x^3}{6} - \dfrac{x^5}{40} \cdots$

4. Suppose $y = \displaystyle\sum_{n=0}^{\infty} c_n x^n$. It then follows that $y'' = \displaystyle\sum_{n=0}^{\infty} n(n-1)c_n x^{n-2} = \displaystyle\sum_{n=0}^{\infty}(n+2)(n+1)c_{n+2}x^n$. The equation $y'' = -y$ results in the recurrence $c_{n+2} = c_n[(n+2)(n+1)]^{-1}$. Moreover $c_0 = y(0) = \cos 0 = 1$ and $c_1 = y'(0) = \sin 0 = 0$.

Exercises 4.2

1. a) $1 + x + x^2$ **c)** $1 + \dfrac{3x}{2} + \dfrac{3x^2}{8}$ **e)** $1 + 2x + 3x^2$

2. a) $0 + 0x + 0x^2$ **c)** $1 - \dfrac{x}{3}$ **e)** $\dfrac{x}{2} + \dfrac{x^2}{8}$ or $2 - \dfrac{x}{2} - \dfrac{x^2}{8}$

g) $1 - \dfrac{x}{5} + \dfrac{2x^2}{125}$ **i)** $2 + \dfrac{x}{25} + \dfrac{2488x^2}{15625}$

4. a) $0 + 0x + 0x^2$ **c)** $1 - \dfrac{x}{3} + \dfrac{x^3}{81} + \dfrac{x^4}{243}$

e) $\dfrac{x}{2} + \dfrac{x^2}{8} + \dfrac{x^3}{16} + \dfrac{5x^4}{128}$ or $2 - \dfrac{x}{2} - \dfrac{x^2}{8} - \dfrac{x^3}{16} - \dfrac{5x^4}{128}$

g) $1 - \dfrac{x}{5} + \dfrac{2x^2}{125} + \dfrac{632x^3}{3125} + \dfrac{118x^4}{15625}$

i) $2 + \dfrac{x}{25} + \dfrac{2488x^2}{15625} + \dfrac{2738x^3}{9765625} + \dfrac{16233997x^4}{1220703125}$

5. a) $2 + \dfrac{3x}{4} + \dfrac{25x^2}{64} + \dfrac{27x^3}{128} + \dfrac{1311x^4}{16384}$ or $-2 + \dfrac{3x}{4} - \dfrac{25x^2}{64} + \dfrac{27x^3}{128} - \dfrac{1311x^4}{16384}$

c) $1 + \dfrac{x}{2} - \dfrac{x^2}{24} - \dfrac{65x^4}{1152}$ or $-1 + \dfrac{x}{2} + \dfrac{x^2}{24} + \dfrac{65x^4}{1152}$

7. a) $1 - \dfrac{x^2}{4} - \dfrac{5x^4}{192}$ **c)** $x + \dfrac{x^3}{3}$ **e)** $x - \dfrac{x^2}{2} + \dfrac{x^3}{3} - \dfrac{x^4}{4}$

Exercises 5.1

4. a) Integrate both sides of the equation of Proposition 5.4 and substitute $x = 0$ in order to get rid of the constant.

b) Integrate the equation of 4a to obtain the equation $\dfrac{x^4}{48} - \dfrac{\pi^2 x^2}{24} + C =$
$\cos x - \dfrac{\cos 2x}{2^4} + \dots$ where $C = 1 - \dfrac{1}{2^4} + \dfrac{1}{3^4} - \dfrac{1}{4^4} \dots$. Then mimic the proof of Proposition 5.1 to show that $C = \dfrac{7\pi^4}{720}$.

7. a) $1 - 1 + 1 - 1 + 1 - 1 \dots = \dfrac{1}{2}$ convergence not valid.

c) $\dfrac{1}{2} + \dfrac{1}{2} - 1 + \dfrac{1}{2} + \dfrac{1}{2} - 1 + \dfrac{1}{2} + \dfrac{1}{2} - 1 \dots = \dfrac{1}{2}$ convergence not valid.

e) $-1 - 1 - 1 - 1 - 1 - 1 \dots = \dfrac{1}{2}$ convergecne not valid.

9. Use $\pi/3$ in Proposition 5.1.3.

11. Use both $\pi/4$ and $3\pi/4$ in Proposition 5.1.3.

13. This follows directly from Exercise 12.

15. Start by substituting $\pi/4$ in Proposition 5.1.4 and also make use of Proposition 5.1.2.

17. $\pi^4/96$

19. $7\pi^2/54$

Exercises 6.1

1. Mimic the proof of 6.1.1 and use the fact that m^2 is divisible by 3 only when m is divisible by 3.

6. Recall that the long division process involves repeated subtractions of the form $r_{k+1} = (10r_k + d_k - e_k \cdot n)$ where d_k is eventually always 0 and e_k is the quotient of $10r_k + d_k$ divided by n. Hence, eventually r_k completely determines both r_{k+1} and e_k. Since r_k is of necessity a positive integer less than n it follows that the value of r_k must evenutally repeat and hence cycle. It follows that the digits e_k will also eventually cycle.

7. This is not a recurrent decimal expansion and so, by Exercise 6, this does not represent a rational number.

9. If $\sqrt{2} + \sqrt{3}$ were rational so would its square be rational. From this would follow the rationality of $\sqrt{6}$, in contradiction with Exercise 6.

Exercises 6.2

1. Use mathematical induction on n.

4. Left-side$= (a+b)c + (a+b)d = ac + bc + ad + bd =$ right-side.

5. Left-side$= a(b + (-c)) = ab + a(-c) =$ right-side.

9. Left-side $= (a+b)^2(a+b)$ and make use of Exercise 7.

11. Use induction to prove that $(1-a)(1 + a + a^2 + \ldots + a^n) = (1 - a^{n+1})$.

12. Add $-a$ to both sides of the equation $a + x = a$.

17. Using previous exercises show that $4a^2 \left(x + \frac{b}{2a}\right)^2 = b^2 - 4ac$.

19. Suppose $\lambda = a/b = c/d$. It follows that $a = b\lambda$ and $c = d\lambda$ so that left-side $= \dfrac{(b\lambda + b)}{(b\lambda - b)} = \dfrac{(\lambda + 1)}{(\lambda - 1)} = (d\lambda + d)/(d\lambda - d) =$ right-side.

23. $a < 2^{-1} \rightarrow 1 - a > 1 - 2^{-1} = 2^{-1} \rightarrow (1-a)^{-1} < 2$

25. Use Proposition 6.2.2.8 and mathematical induction.

27. Use Proposition 6.2.2.10 and proof by contradition.

29. Cross multiply.

31. Use Exercise 30.

33. Use Exercise 30.

34. a) Observe that $a = (a+b) + (-b)$ and use the triangle inequality.

35. Show that if such an ordering existed then you could derive a contradiction out of each of the inequalities $\sqrt{-1} \geq 0$ and $\sqrt{-1} \leq 0$. Hence, by O1, no such ordering can exist.

40. a) $\{0,1,2,3,4,5,6,8\}$ c) $\{0,1,\ldots,9\}$ e) $\{1,3,5\}$ g) $\{1,3,5\}$
i) $\{2,4,6\}$ k) $\{2,4,6,8\}$ m) 1   o) 0

42. The numbers $1/n, 2/n, \ldots, (n-1)/n$ divide the interval $[0,1]$ into n parts. Hence one of these parts must contains two of the $n+1$ numbers $0, 1, x - [x], 2x - [2x], \ldots, (n-1)x - [(n-1)x]$. The subtraction of these two numbers yields the required result.

Exercises 6.3

1. a) $1 \in S, 0 \notin S$ c) $10 \notin S, -1 \in S$ e) $\sqrt{2} \notin S$, no g.l.b.
g) no l.u.b., $0 \notin S$,   i) $0 \in S$, no g.l.b.   k) no l.u.b., $0 \notin S$
m) $0 \in S$, no g.l.b.   o) $\sqrt[3]{17} \notin S, -\sqrt{3} \notin S$ q) $1/9 \notin S, .1 \in S$
s) $1 \notin S, 0 \in S$ u) $3/2 \in S, 0 \notin S$.

3. Use the fact that $\left(\sqrt{a} - \sqrt{b}\right)^2 \geq 0$

5. Suppose not, then there are two distinct positive numbers a and b such that $b^n - a^n = 0$. Use part 10 of Proposition 6.2.2 to derive a contradiction.

7. Set $S = \{u \in \mathbb{R} | (-\infty, u) \subset M\}$. Since S is bounded above and non empty, it has a least upper bound U. If $a < U$, then a is not an upper bound of S and so there exists $b > a$ such that $a \in (-\infty, b) \subset M$ and hence $(-\infty, U) \subset M$. If $(-\infty, r) \subset M$ then $r \in S$ and so $r \leq U$.

Exercises 6.4

1. $\{(1,37),(0,1)\}$

3. $\{(1/3, 1/2), (1/6, 1/3), (0, 1/6)\}$

5. $\left(\sqrt{3} - 1, 1\right), \left(2 - \sqrt{3}, \sqrt{3} - 1\right)$

8. Repeat the proof of Proposition 6.4.5. using $\sqrt{n^2 + 1} \pm n$ instead of $\sqrt{2} \pm 1$.

10. First prove that if $E(a, b) = (r, a)$ then every common divisor of a and b is also a common divisor of r and a. From this conclude that (a, b) and (r, a) have the same greatest common divisors. Now apply this to the the sequence $E(s, t)$.

11. 18

13. 3

Exercises 6.5

1. Let $c = f(1)$. Then prove by induction that $f(n) = cn$ for positive integers. Next show this also holds for negative integers. Now show that $f(r) = cr$ whenever $r = 1/n$ for some positive integer. Conclude that $f(r) = cr$ for every rational number.

3. For all x : $((f \circ g) \circ h)(x) = (f \circ g)(h(x)) = f(g(h(x))) = f((g \circ h)(x)) = (f \circ (g \circ h))(x)$. Hence $(f \circ g) \circ h = f \circ (g \circ h)$.

5. No. Use $f(x) = x^2$ and $g(x) = h(x) = x$ and any nonzero value of x.

6. Suppose $a \neq b$. Then $g(a) \neq g(b)$ because g is one-to-one and $f(g(a)) \neq f(g(b))$ because f is one-to-one. Consequently $(f \circ g)(a) \neq (f \circ g)(b)$ and so $f \circ g$ is also one-to-one.

Exercises 7.1

3. $\{(-1)^n\}$

5. Fix m and let $\varepsilon > 0$. For every $n \geq \max\{2/\varepsilon, 2m\}$ we then have $|a_n - a_m| \leq 1/(n - m) \leq n/2 \leq \varepsilon$ so that $\{a_n\} \to a_m$. It follows from Exercise 4 that all the a_m's are equal.

7. $\{a_0, b_0, a_1, b_1, a_2, \}$, $\{1, a_0, b_0, a_1, b_1, \dots\}$, etc.

10. For each positive integer m, let L_m be the finite list $a_{0,m}, a_{1,m-1}, a_{2,m-2}, a_{3,m-3}, \dots, a_{m,0}$. Then the sequence $\{L_0, L_1, L_2, \dots\} = \{a_{0,0}, a_{0,1}, a_{1,0}, a_{0,2}, a_{1,1}, a_{2,0}, \dots, a_{0,m}, a_{1,m-1}, a_{2,m-2}, a_{3,m-3}, \dots, a_{m,0}, \dots\}$ formed by concatenating all these lists contains all the terms $a_{k,n}$.

11. Use Exercise 10.

14. For each positive integer m let L_m be a finite list consisting of all the solutions to equations of the form $ax^2 + bx + c = 0$ where $|a| + |b| + |c| = m$. Now proceed as in Exercise 10.

16. Suppose $\{x_n\}$ is a sequence that contains all the real numbers. For each m let d_m be a digit that is different from the digit in the nth decimal place of the decimal expansion of x_m (and is also different from 9). Let $d = .d_1 d_2 d_3 \dots$. The avoidance of 9's guarantees that d has a unique decimal expansion. By construction $d \neq x_n$ for each n, contradicting the fact that $\{x_n\}$ contains all the real numbers.

Exercises 7.2

1. a) Diverges because it is unbounded: $a_n > n$.

c) Converges to 0   e) Converges to 0   g) Converges to 1

i) Converges to 0   k) Converges to 0   m) Converges to 0

because it can be proved by induction that $n^2 < 2^n$ for $n \geq 5$.

o) Converges to 0. Use Exercise 6.3.3 to prove that $n! \leq \left(\dfrac{n}{2}\right)^n$.

3. See the trick of Example 7.2.17.

5. Show that the sequence $\{1/a_n\}$ is unbounded.

7. $\displaystyle\lim_{n\to\infty} b_n = \frac{1}{2}\left(\lim_{n\to\infty} a_n + \lim_{n\to\infty} a_{n+1}\right) = \frac{1}{2}(A + A) = A$

9. Since the sequence $\{a_n\}$ is bounded, $\{b_n\} \to 0$.

11. Depends. If $a_n = 1/n$, then $\{b_n\}$ diverges, but if $a_n = 1/n^3$, then $\{b_n\}$ converges.

13. $a_n = (-1)^n$.

17. 1) Proceed by contradiction.   2) Apply part 1 to $\{a_n - b_n\}$.

19. Assume that $A = 1, a_n \geq 1$ for all n, and $r = 1/k$. Then

$$\left|a_n^{1/k} - 1\right| = \frac{|a_n - 1|}{a_n^{(k-1)/k} + a_n^{(k-2)/k} + \ldots + 1} \leq \frac{|a_n - 1|}{n}$$

and so $a_n^r \to 1 = A^r$.

If r is an integer, then $a_n^r \to A^r$ because of Theorems 7.2.3.3, 5. Hence the desired result follows for any rational r.

If $A = 1$ and $a_n \leq 1$ for all n, then by the above $\{1/a_n^r)\} \to 1$ and so again $a_n^r \to 1 = A^r$. It now follows from Proposition 7.2.11 that the desired conclusion holds regardless of the value of a_n. Finally, if $A \neq 1$, apply the above to the sequence $b_n = a_n/A$.

23. Use Proposition 6.3.6.

29. It is easy to find such sequences where $A = B = 0$.

31. Make use of Exercise 14 and the squeezing principle.

33. $\displaystyle\lim_{n\to\infty} \frac{mx}{m+n} = 0$ and $\displaystyle\lim_{m\to\infty} \frac{mx}{m+n} = x$.

35. Note that $a_{n+1} + 1 = (a_n + 1)/2$ and then use Propositions 7.2.7 and 7.2.8.

Exercises 8.1

1. Diverges

3. Limit $= \sqrt{(1 + \sqrt{5})/2}$

5. Limit $=$ solution of $A^3 = A^2 + 1$ which is $1.46557\ldots$

7. Hint: Prove by induction that $1 \le a_n < 3$. Limit $= \left(3 + \sqrt{5}\right)/2$

9. Limit $= 1$

11. Limit $= 1$

13. Prove by induction and cross multiplication that $a_{n+1} > a_n$.
Limit $= \left(\sqrt{5} - 1\right)/2$.

15. Diverges

19. Show that $a_{n+1} < a_n$ iff $a_n < a_{n-1}$ and apply Exercise 17 to prove convergence.

21. See hint for 19

23. Converges

25. Diverges

27. See Example 8.1.3

33. See hint for 29

35. See Example 8.1.1

37. Prove by induction that $a_{n+1} < a_n$ iff $a_n > a_{n-1}$ and apply Exercise 17

39. $f(x) = 2x + 1$

43. Make use of Exercise 42.

45. This series converges because $a_2 = 2$ and hence $a_n \le 1 + 1 + 1/2 + 1/2^2 + 1/2^3 + \ldots < 3$.

Exercises 8.2

1. Note $|a_{n+m} - a_n| \le |a_{n+m} - a_{n+m-1}| + |a_{n+m-1} - a_{n+m-2}| + \ldots + |a_{n+1} - a_n|$ and use Proposition 1.1.2 to prove that $\{a_n\}$ has the Cauchy Property.

5. a) $|(a_{n+m} + b_{n+m}) - (a_n + b_n)| \le |a_{n+m} - a_n| + |b_{n+m} - b_n|$.
d) $|a_{n+m}b_{n+m} - a_n b_n| \le |a_{n+m} - a_n||b_{n+m}| + |a_n||b_{n+m} - b_n|$ and proceed in a manner similar to that in the proof of Theorem 7.2.3.3.

7. First show that there is a number M such that $|a_{n+1} - a_n| < Mc^n$. Then show that $|a_{n+m} - a_n| < Mc^n/(1 - c)$.

Exercises 9.1

1. a) 3   **c)** 11/8   **e)** 7/2   **g)** 51

2. a) Diverges   **c)** Diverges   **e)** Diverges   **g)** Converges

3. Use Exercise 6.2.42.

7. Use an integral comparison.

9. See Exercise 3 above.

Exercises 9.2

1. a) Diverges   **c)** Converges

e) Converges. Compare to $\displaystyle\int_2^\infty \frac{dx}{(\ln x)^{\ln x}}$ and substitute $u = \ln x$.

g) Diverges   **i)** Converges. Show first that $n < 2^n$ **k)** Diverges

m) Diverges   **o)** Diverges   **q)** Converges

3. The given convergence implies that $a_n \to 0$ and hence it may be assumed that $0 \le a_n < 1$.

5. For any positive integer k, the number of k-digit integers that are 1-free is $8 \cdot 9^{k-1}$, and the reciprocal of each is at most $10^{-(k-1)}$. Hence the given series is less than the convergent series $\sum_{k=1}^\infty 8 \cdot (9/10)^{k-1}$.

8. Note that $0 \le c_n \le a_n + b_n$.

9. Let d_n, e_n, f_n be the partial sums of $\sum a_n, \sum b_n, \sum c_n$ respectively. Show that $d_n e_n \le f_{2n} \le d_{2n} e_{2n}$ and apply the squeezing principle.

11. Yes.

Exercises 9.3

1. a) Diverges   **c)** Converges   **e)** Diverges   **g)** Diverges

i) Diverges   **k)** Diverges   **m)** Converges   **o)** Converges

q) Converges

3. By Exercise 6.2.30, $0 \le |a_n b_n| \le (a_n^2 + b_n^2)/2$.

5. $\displaystyle\sum_{n=0}^{2m}(-1)^n a_n) = (1 + 1/2 + 1/3 + \ldots + 1/(m+1)) - (1 + 1/2 + \ldots + 1/2^{m-1})$
$\geq (1 + 1/2 + 1/3 + \ldots + 1/(m+1)) - 2.$

7. Let M be a bound of $\{b_n\}$. Use the convergence of $\sum M a_n$ to prove that $\sum a_n b_n$ has the Cauchy Property.

9. Prove that $\sum c_n$ has the Cauchy Property.

11. See Example 9.3.7

13. This is clear for $x = 0, 1$. For other values of x use the ratio test.

15. Set $a_n = \dfrac{1}{2n+1} + \dfrac{1}{2n+2}$ and use Proposition 9.3.8.

17. No.

19. Use induction.

21. No.

Exercises 9.4

1. Also make use of the Fundamental Theorem of Calculus.

2. For any prime p, the probability that a randomly selected integer is not divisible by p^2 is $1 - 1/p^2$.

Exercises 10.1

1. a) $[-1, 1)$ **c)** $[-1, 1)$ **e)** $(-1, 1)$ **g)** $(-\infty, \infty)$ **i)** $[-1, 1)$
k) $(-1, 1]$ **m)** $(-1, 1)$ **o)** $(-a^{-1}, a^{-1})$ **q)** $(-1, 1)$ **s)** $(-1, 1)$
u) $(-1, 1)$ **w)** $\{0\}$

3. $\sum(-1)^n x^n/n$

5. Use Propositions 9.1.2 and 9.3.8.

7. No.

9. Use Theorem 9.3.8 and the fact that $\dfrac{n^n}{(n+1)^{n+1}} < 1/n.$

11. $\sum(-1)^n x^n/(na^n)$

13. $\sum x^n/(a^n n^2)$

15. Suppose $|x| < \rho_c$. Then $\sum c_n x^n$ converges absolutely and so $\sum a_n x^n$ converges absolutely so that $|x| \leq \rho_a$. Consequently $\rho_a \geq \rho_c$. A counterexample for the second part is provided by $c_n = n^{-1/2}$ and $a_n = (-1)^n n^{-1/2}$.

19. This series converges for all x in $(-1, 1)$ and nowhere else.

21. This series converges for all x in $(-1/8, 1/8)$ and nowhere else.

Exercises 10.2

1. a) No   c) Yes   e) No   g) No   i) No

2. a) $\alpha > 1$ c) $\alpha > 4$ e) $\alpha > -1$ g) All α i) All α

5. $a_n / a_n \to 1$.

7. $n = O(n^2)$ but $n^2 \neq O(n)$.

9. $n = O(n^2), 1/n \neq O(1/n^2)$

Exercises 11.2

1. a) 10   c) No limit   e) No limit   g) $-1/5$ i) 0   k) 0
m) 1/2

3. a) 0 c) No limit

5. $f(x) = 0$ if $x < 2$ and $f(x) = 1$ if $x \geq 2$

7. $f(x) = 0$ if $x < -1$, $f(x) = 1$ if $-1 \leq x < 0$, $f(x) = 0$ if $0 \leq x < 1$ and $f(x) = 1$ if $x \geq 2$.

13. Let $x_n \to a$. Then for sufficiently large n, $|f(x_n)| > |A/2|$ and $|g(x_n)| < |A|/(2k)$, where k is any positive integer. Conclude that $\{f(x_n)/g(x_n)\}$ is an unbounded sequence.

Exercises 11.3

1. a) Continuous with domain $x \neq \pm 5$
    c) Continuous with domain $\mathbb{R}$
    e) Discontinuous only at $x = 5$
    g) Continuous in domain $x \neq 0$
    i) Continuous everywhere
    k) Continuous with domain $x \neq 0$
    m) Continuous everywhere

3. Yes

5. No

7. Yes

9. Yes

11. No

13. Yes

15. Given any sequence $\{x_n\} \to a$ the neighborhood $(a - \delta, a + \delta)$ contains almost all of its terms.

17. Show that if x is an irrational number and $\left\{ \dfrac{p_n}{q_n} \right\}$ is a sequence of rational numbers that converge to x, then for every $\varepsilon > 0$ $|q_n| > 1/\varepsilon$ for almost all n.

19. f is continuous only at 0.

21. Use $\cos \alpha - \cos \beta = -2 \sin \frac{\alpha+\beta}{2} \sin \frac{\alpha-\beta}{2}$.

23. Use $\tan x = \sin x / \cos x$

25. a) Continuous   c) Continuous

31. This follows from Example 11.3.9 and Proposition 11.3.4

33. Fix x and define $x_n = x/2^n$ for $n = 0, 1, 2, \ldots$. Then $f(x) = f(x_0) = f(x_1) = f(x_2) = \ldots$. Hence $f(0) = \lim_{n \to \infty} f(x_n) = f(x)$.

35. If $0 < x < 1$ then $f(x) = f\left(x^2\right) = f\left(x^4\right) = f\left(x^8\right) = \ldots = f(0)$.
Also $f(0) = f(x) = f\left(x^{1/2}\right) = f\left(x^{1/4}\right) = f\left(x^{1/8}\right) = \ldots = f(1)$.
If $x > 1$ then $f(x) = f\left(x^{1/2}\right) = f\left(x^{1/4}\right) = f\left(x^{1/8}\right) = \ldots = f(1) = f(0)$.
Finally, if $x < 0$, then $f(x) = f\left(x^2\right) = f(-x) = f(0)$.

37. Suppose not. Then for every positive integer n there is a number $x_n \in (x^* - 1/n, x^* + 1/n)$ such that $f(x_n) = 0$. Since $\{x_n\} \to x^*$ it follows from the continuity of f at x^* that $f(x^*) = 0$ which is the required contradiction.

Exercises 11.4

1. Define $F(x) = f(x) - x$ and apply the Intermediate Value Theorem to $F(x)$.

7. Define $f(x) = x^4 - 5x^3 + x - 1$ and observe that $f(0) = -1$ and $f(-1) = 4$.

9. Apply the I.V.T. to $f(x) = 2 \sin x - \cos x - 1 + x^2$.

15. Let $a < b$ be such that $f(a) = f(b) = 0$ and apply the maximum principle to obtain a contradiction.

Exercises 12.2

3. If $r = m/n$ where m is any integer and n is a positive integer note that x^r is the composition of the two function $x^{1/n}$ and x^m

5. $\tan x = \sin x / \cos x$

7. Treat the two cases $x \geq 0$ and $x \leq 0$ separately.

9. No

11. No

13. No

15. yes

17. Observe that $(f(x + h) - f(x))/h \geq 0$.

19. $x^2 + 3x + 9$

21. Note that $\dfrac{f(ax) - f(bx)}{cx} = \dfrac{a}{c} \cdot \dfrac{f(ax) - f(0)}{ax} - \dfrac{b}{c} \cdot \dfrac{f(bx) - f(0)}{bx}$.

23. Note that $f(a + h) - f(a - h) = [f(a + h) - f(a)] - [f(a - h) - f(a)]$.
Use $f(x) = x/x$.

25. $\left| \dfrac{f(a + h) - f(a)}{h} \right| \leq ch$

27. $\dfrac{f(a + h) - f(a)}{h} = \dfrac{A + h \cdot e(a + h) - A}{h} = e(a + h)$.

29. Show that $F(a + h) - F(a) = h \cdot f(a + h)$.

31. Treat the cases $x > 0, x < 0, x = 0$ separately.

Exercises 12.3

3. This is similar to Example 12.3.4

5. Apply the MVT to $f(x)$ on $[x, 0]$ or $[0, x]$, whichever is applicable.

7. Apply the MVT to $f(x) = x^n$ on $[x, y]$.

9. Apply the MVT to $f(x) = x^r$ on $[1, 1 + h]$. Also true for $h > -1$.

11. Apply the MVT to $f(x) = \ln(1 + x)$ on $[0, x]$ or $[x, 0]$.

13. Apply the MVT to $f(x)$ on $[0, x]$ to conclude that $xf'(x) - f(x) > 0$.
Show that $f(x)/x$ has a nonnegative derivative.

15. See Exercise 12.2.25

Exercises 12.4

1. Let $F(x)$ and $G(x)$ be antiderivatives of $f(x)$ and $g(x)$ respectively. Then $AF(x) + BG(x)$ is an antiderivative of $Af(x) + Bg(x)$.

3. Note that $-|f(x)| \le f(x) \le |f(x)|$ and apply Exercises 1 and 2.

5. If $F(x)$ is any antiderivative of $f(x)$ then $F(c) - F(a) = F(b) - F(a) + F(c) - F(b)$.

Exercises 13.1

1. Does not converge outside $[-1/2, 1/2]$

3. Converges but not uniformly.

5. Converges but not uniformly

7. Converges uniformly

9. Converges uniformly

11. Does not converge for $x \ge 0$.

13. Converges uniformly

15. Converges uniformly

17. Does not converge for any x

19. Converges uniformly

21. If $|t_m(x)| < \varepsilon$ for all $m > n'$ and $x \in D$ and $|t_m(x)| < \varepsilon$ for all $m > n''$ and $x \in E$, then $|t_m(x)| < \varepsilon$ for all $m > \max\{n', n''\}$ and $x \in D \cup E$.

23. Use the fact that $\left| \sum_{n=m}^{\infty} (g_n(x) + h_n(x)) \right| \le \left| \sum_{n=m}^{\infty} g_n(x) \right| + \left| \sum_{n=m}^{\infty} h_n(x) \right|$

25. No. Take $g_n(x) = h_n(x) = (-1)^n/\sqrt{n}$ for a counterexample.

27. Let k be any positive integer. If $m > b + k$, then $|t_m(x)| < \sum_{n=k}^{\infty} e^{-n^2} < \sum_{n=k}^{\infty} e^{-n} < e^{-k+1}$. The series does not converge uniformly in $\mathbb{R}$.

29. No. Use $g_n(x) = (-1)^n/n$ as a counterexample.

33. $\sum g_n(x) = \sum x^n - \sum (x^2)^n = 1/(1-x) + 1/(1-x^2)$. $\sum g_n(1) = \sum 0 = 0$ but $\lim_{x \to 1} x/(1 - x^2)$ does not exist.

35. Show that $t_m(x) = (1 + x^2)^{-m}$.

37. Show that $1/(x^2 + n) - 1/(x^2 + n + 1) < 1/n(n+1) = 1/n - 1/(n+1)$. This convergence is never absolute.

41. a) No, no, yes, yes   c) Yes[4]   e) Yes, no, yes, yes   g) No, no, yes, yes   i) No[4]   k) Yes[4]

43. Uniform

45. Not uniform

47. Uniform

Exercises 13.2

1. Not enough information

3. True

5. True

7. Not enough information

9. True

11. False; this series does not converge for $|x| > 1$

13. True

15. False; the series does not converge for $x > 1$.

17. True

19. Use Exercise 18 to show that the coefficients of the series $\sum (a_n - b_n)x^n$ must all be zero.

21. Apply Theorem 13.2.4 to the function $f(x) \sin kx$

25. $\left| \sum\limits_{n=m}^{\infty} \int_a^x g_n(t)dt \right| = \left| \int_a^x \sum\limits_{n=m}^{\infty} g_n(t)dt \right| \leq \int_a^b \left| \sum\limits_{n=m}^{\infty} g_n(t) \right| dt$

27. Any such series would have to be differentiable at 0.

29. a) $x(1 + x)(1 - x)^{-3}$ b) 6

31. $t_m(x) = x^m$

33. The limit function is not continuous.

Exercises 14.1

1. Integrate the equation of 14.1.4. You will need Exercise 13.2.25.

3. a) Combine Exercise 2 with Proposition 14.1.1.
   b) Mimic the proof of Proposition 14.1.2.
   c) $\int_a^b (\cos\phi + \cos 3\phi + \ldots)d\phi = 0$ as long as $\pm\pi, 0 \notin [a, b]$.
   d) Use Proposition 14.1.3 e) Treat $x > 0$ and $x < 0$ separately.

Exercises 14.2

1. Mimic the proof of Proposition 14.2.3.

3. Mimic the proof of Proposition 14.2.3.

5. Using the notation of 14.2.1, let $d \in (a, c)$ be such that $f''(d) = (f'(c) - f'(a))/(c - a)$. Then $|f(x_M)| = |f'(c)| \cdot |x_M - a| = |f''(d)| \cdot |c - a| \cdot |x_M - a| \le |f(x_M)| \cdot |c - a| \cdot |x_M - a|$

7. a) If y, z are two solutions on any $[a, b]$, then $|y' - z'| = |y - z|$.
   c) If y, z are two solutions on any $[a, b]$, then $|y' - z'| = |x| \cdot |y - z| \le A|y - z|$, where $A = \max\{|a|, |b|\}$.
   e) If y, z are two solutions on $(-1, 1)$, then $|y' - z'| = |2x| \cdot |y - z| \le 2|y - z|$.
   g) If y, z are two solutions on any $[a, b], 0 < a < b < \infty$, then $x^2|y' - z'| = |y - z|$ and so $|y' - z'| \le a^{-2}|y - z|$.

Bibliography

Baron, M. E. (1987). *The Origins of the Infinitesimal Calculus*. Dover (reprint), New York.

Bartle, R. G., and Sherbert, D. R. (1992). *Introduction to Real Analysis*, second edition. John Wiley & Sons, Inc., New York.

Birkhoff, G. (1973). *A Source Book in Classical Analysis*. Harvard University Press, Cambridge.

Bottazini, U. (1986). *The Higher Calculus: A History of Real and Complex Analysis from Euler to Weierstrass*. Springer-Verlag, New York.

Boyer, C. B. (1959). *The History of Calculus and Its Concpetual Foundations*. Dover (reprint), New York.

Bressoud, D. (1994). *A Radical Approach to Real Analysis*. The Mathematical Association of America, Washington, D.C.

Bromwich, T. (1947). *Introduction to the Theory of Infinite Series*, second edition. MacMillan & Co., London.

Calinger, R. (1982). *Classics of Mathematics*. Moore Publishing Co., Oak Park, Illinois.

Cauchy, A.-L. (1899). *Cours d'Analyse*, Paris, 1821. In *Ouevres Complètes, d'Augustin Cauchy*, series 2, vol. 3, Guathiers-Villars, Paris.

Courant, R. (1959). *Differential and Integral Calculus*. Interscience Publishers, Inc., New York.

Davis, P. J., and Hersh, R. (1981). *The Mathematical Experience*. Houghton Mifflin Co., Boston.

Descartes, R. (1954). *The Geometry*. Dover, New York.

Edwards, C. H. Jr. (1979). *The Historical Development of the Calculus*. Springer-Verlag, New York.

Edwards, H. M. (1974). *Riemann's Zeta Function*. Academic Press, New York.

Euler, L. (1760). Variae observationes circa series infinitas. In *Opera*, **14**(1):216–244.

Euler, L. (1754–55). Subsiduum calculi sinuum. *Novi commentarii academiae scientarium Petropolitanae*, **5**:164–204; published in *Opera*, **14**(1):542–584.

Ewald, W. B. (1996). *From Kant to Hilbert, A Source Book in the Foundations of Mathematics*. Clarendon Press, Oxford.

Fauvel, J. and Gray, J. (1987). *The History of Mathematics, A Reader*. MacMillan Education, London.

Grabiner, J. V. (1981). *The Origins of Cauchy's Rigorous Calculus*. MIT Press, Cambridge, Massachussetts.

Grabiner, J. V. (1983). Who gave you the epsilon? Cauchy and the origins of rigorous calculus. *Amer. Math. Monthly*, **90**:185–194.

Grattan-Guinness, I. (1970). *The Development of the Foundations of Mathematical Analysis from Euler to Riemann*. The MIT Press, Cambridge, Massachussetts.

Grattan-Guinness, I., ed., (1980). *From the Calculus to Set Theory, 1630–1910*. Duckworth, London.

Hairer, E., and Wanner, G. (1996). *Analysis by its History*. Springer-Verlag (UTM Reading in Mathematics), New York.

Hardy, G. H. (1918). Sir George Stokes and the concept of uniform convergence. *Proc. Camb. Phil. Soc.*, 19:148–156.

Lagrange, J. L. (1813). *Theorie des Fonctions Analytique*. Courcier, Paris.

Landau, E. (1951). *The Foundations of Analysis*, Chelsea Publishing Co., New York.

Newton, I. (1964). *The Mathematical Works of Isaac Newton* (D. T. White-side, ed.) Johnson Reprint Corporation, New York.

Newton, I. (1967). *The Mathematical Papers of Isaac Newton* (D. T. White-side, ed.) Cambridge University Press, London.

Riemann, B. (1953). *Gesammelte Mathematische Werke.* Dover, New York.

Ross, K. A. (1980). *Elementary Analysis: The Theory of Calculus.* Undergraduate Texts in Mathematics, Springer-Verlag, New York.

Rudin, W. (1964). *Principles of Mathematical Analysis.* McGraw-Hill, New York.

Rüthing, D. (1984). Some definitions of the concept of function from Joh. Bernoulli to N. Bourbaki. *Mathematical Intelligencer,* 6:72–77.

Siu, M.-K. (1995). Concept of a function—its history and teaching. In *Learn from the Masters* (F. Swetz, J. Fauvel, O. Bekken, B. Johansson, and V. Katz, eds.) Mathematical Association of America, Washington, D.C.

Smith, D. E. (1959). *A Source Book in Mathematics.* Dover, New York.

Struik, D. J. (1986). *A Source Book in Mathematics, 1200–1800.* Princeton University Press, Princeton, New Jersey.

Weierstrass, K., Differential Rechnung, Vorlesung and dem Königlichen Gewerbeinstitute, (manuscript 1861, typewritten by H. A. Schwarz) Math. Bibl. Humboldt Universität, Berlin.

Index